Lecture Notes in Networks and Systems

Volume 303

The series "Lecture Notes in Networks and Systems" publishes the latest developments in Networks and Systems—quickly, informally and with high quality. Original research reported in proceedings and post-proceedings represents the core of LNNS.

Volumes published in LNNS embrace all aspects and subfields of, as well as new challenges in, Networks and Systems.

The series contains proceedings and edited volumes in systems and networks, spanning the areas of Cyber-Physical Systems, Autonomous Systems, Sensor Networks, Control Systems, Energy Systems, Automotive Systems, Biological Systems, Vehicular Networking and Connected Vehicles, Aerospace Systems, Automation, Manufacturing, Smart Grids, Nonlinear Systems, Power Systems, Robotics, Social Systems, Economic Systems and other. Of particular value to both the contributors and the readership are the short publication timeframe and the world-wide distribution and exposure which enable both a wide and rapid dissemination of research output.

The series covers the theory, applications, and perspectives on the state of the art and future developments relevant to systems and networks, decision making, control, complex processes and related areas, as embedded in the fields of interdisciplinary and applied sciences, engineering, computer science, physics, economics, social, and life sciences, as well as the paradigms and methodologies behind them.

Indexed by SCOPUS, INSPEC, WTI Frankfurt eG, zbMATH, SCImago.

All books published in the series are submitted for consideration in Web of Science.

More information about this series at http://www.springer.com/series/15179

Lalit Garg · Nishtha Kesswani ·
Joseph G. Vella · Peter A. Xuereb ·
Man Fung Lo · Rowell Diaz ·
Sanjay Misra · Vipul Gupta ·
Princy Randhawa
Editors

Information Systems and Management Science

Conference Proceedings of 3rd International Conference on Information Systems and Management Science (ISMS) 2020

Springer

Editors
Lalit Garg
CIS Dept
University of Malta
Msida, Malta

Nishtha Kesswani
Department of Computer Science
Central University of Rajasthan
Rajasthan, India

Joseph G. Vella
CIS Dept
University of Malta
Msida, Malta

Peter A. Xuereb
Faculty of ICT
University of Malta
Msida, Malta

Man Fung Lo
The Education University of Hong Kong
Hong Kong, Hong Kong

Rowell Diaz
Nueva Ecija University of Science
and Technology
Cabanatuan City, Philippines

Sanjay Misra
Computer Engg, CUCRID Bldg, #A301
Covenant University
Ota, Nigeria

Vipul Gupta
Lm Thapar School of Management
Thapar University
Mohali, India

Princy Randhawa
Manipal University Jaipur
Jaipur, India

ISSN 2367-3370 ISSN 2367-3389 (electronic)
Lecture Notes in Networks and Systems
ISBN 978-3-030-86222-0 ISBN 978-3-030-86223-7 (eBook)
https://doi.org/10.1007/978-3-030-86223-7

This Springer imprint is published by the registered company Springer Nature Switzerland AG
The registered company address is: Gewerbestrasse 11, 6330 Cham, Switzerland

Preface

We are pleased to present this volume of the selected contributed articles, submitted and presented in the 3rd International Conference on Information Systems and Management Science (ISMS) 2020, which held on December 15 and 16, 2020, at The faculty of ICT, University of Malta, Msida, Malta, in collaboration with the International Association of Academicians (IAASSE), USA. (Conference link: http://isms.iaasse.org/.). The 3rd International Conference on Information Systems and Management Science (ISMS 2020) is a platform for researchers, engineers, academicians as well as industrial professionals from all over the world to present their research results and development activities in information systems and management sciences. The event (ISMS 2020) took place in hybrid model, thereby allowing some authors to present their papers virtually because of COVID-19 and the government's restriction on any gathering. And by that, organizers fulfilled their commitments with researchers and supports the scientific research.

The papers presented in the ISMS 2020 went through strict refereeing and examination resulting in a current acceptance rate of 48.9%. All papers were selected for oral presentation in the conference after an initial review. Every submitted full length paper was sent for peer review to at least two potential reviewers of related areas of expertise once it was passed by the program committee. We are delighted to say that this is in no small part due to the hard work the editorial board and reviewers, in not only refereeing the papers submitted but raising the standard of the quality of papers that are to be published. Last but not least, as the guest editors of ISMS 2020 proceedings, we are thankful to people who have worked with us in planning and organizing both technical arrangements. We also thank all learned authors for their kind cooperation and contribution. Hope that the quality research work published in this issue will be able to serve something new to Humanity and Science.

January 2021

Lalit Garg
Nishtha Kesswani
Joseph G. Vella
Peter Xuereb
Man Fung Lo
Rowell Diaz
Sanjay Misra
Vipul Gupta
Princy Randhawa

Organization

Program Chair

Lalit Garg	University of Malta, Malta

Program Committee

Peter Xuereb	University of Malta
Sanjay Misra	Covenant University, Nigeria
Vipul Gupta	Thapar University, India
Santosh Patil	Manipal University Jaipur, India
Princy Randhawa	Manipal University Jaipur, India
Nishtha Kesswani	Central University of Rajasthan, India
Shahbaz Siddiqui	Manipal University Jaipur, India
Joseph G Vella	University of Malta, Malta
Mohit Jain	Manipal University Jaipur, India
Lalit Garg	University of Malta
Man Fung Lo	The Education University of Hong Kong, Hong Kong
Rowell Diaz	Nueva Ecija University of Science and Technology, Philippines

Contents

Creating High-Performance Work System in Technology Startups

Prayag Gokhale[1(✉)] and Praveen M. Kulkarni[2]

[1] KLE Dr. M. S. Sheshgiri College of Engineering and Technology, Udyambag, Belagavi 590008, Karnataka, India
prayaggokhale@klescet.ac.in
[2] KLS Gogte Institute of Technology, Belagavi 590008, Karnataka, India
pmkulkarni@git.edu

Abstract. Modern manufacturing provides an opportunity to invest in the latest technologies and systems for improving productivity. Innovative technologies and systems such as the High-performance work system enhance the skills of the employees. This study is undertaken to understand the relationship between objectives of the implementation of High-performance works system and its application in technology startups, and to gain a deeper understanding of the implementation of High-performance works system in technology startups further, the study also considers the role of managerial capabilities required for the effective implementation of a High performance works system in technology startups. The data used for the study was collected from 28 Technology start-ups in Karnataka, India, using a structured questionnaire. The results have indicated that the role of managerial capabilities and clarity with regards to objectives of the implementation of a High-performance works system is paramount for effective implementation in technology startups.

Keywords: High-performance work system · Technology startups · Managerial capabilities · Human resource management · Manufacturing capabilities · Competency

1 Introduction

High-performance works systems are a block of HRM practices that characteristically consist of the following emphases: staffing, training, flexible work assignments, communication, decentralized decision making, self-management teams, and compensation [1]. A high-performance works system is influencing the technology startups, for realizing the full potential of a High-performance works system requires an integration of managerial capabilities and manufacturing capabilities. One of the most common factors that influence realizing managerial capabilities and manufacturing capabilities is the investment in this Human resource and technology.

Technology startups have become the foundation for economic growth in today's world of business. Yet only a few factors are known that contribute to the success or

L. Garg et al. (Eds.): ISMS 2020, LNNS 303, pp. 1–10, 2022.
https://doi.org/10.1007/978-3-030-86223-7_1

failure of a startup [2]. Most of the research work done in the area of High-Performance Work System attempts toward the study of direct association amongst the management practices and the Performance outcomes [3]. This research study reports this expansive inquiry by exploring the role that high-performance work systems (HPWS) with special consideration to managerial capabilities and clarity and the objectives of the organization.

A high-performance work system comprises of the most important task of Staffing, which take into consideration the courses of actions wherein the capabilities for job fit and organization fit are appraised. The comprehensiveness of this procedure is based on the evaluations at three different levels. These assessments are established on basis of "knowledge, skills, and abilities (KSAs)", which ultimately result in selecting the best candidate for the specified position. The technology Startups do not have time to undergo a very detailed staffing process [4] whereas it is argued that staffing is the most important portion of HPWS which enables the organization in achieving competitive advantage. Highest, KSA value and distinctiveness is conceivably not essential for every single position, but the significance is in determining the appropriate Job fit and organization fit to augment augmented performance of both organization and individuals. Finally, as a prologue to careful staffing, thoughtfulness in attracting the right applicants from an organizational level [5] may be a significant and disheveled characteristic to confirm selection from the best talent pools.

2 Literature Review

In this section, the study presents a literature review on the application of a High-performance works system, managerial capabilities related to a High-performance works system, and factors associated with objectives of the implementation of a High-performance works system in SMEs.

2.1 High-Performance Work System

The theory of a high-performance work system (HPWS) creates a prerogative that there subsists a structure of work practices for basic workforces in a business that points in some manner to superior performance [5]. Both high-performance work systems and technology Startups have established increasing consideration in HRM research. Conversely, the works on HRM in technology Startups have largely fixated on the problems of "homogeneity" versus "heterogeneity" behaviours, on the one hand, and the opposite amongst the "small is beautiful" and the "bleak-house" perspectives on the other [6]. The stakeholder associations and the assurance of owners to human resource management (HRM) are the two key facets defining the implementation of HPWS [7]. The basic social structure of an organization can arbitrate the relationship between high-performance work systems (HPWS) and organizational performance. HPWS impacts the core social structure by expediting associating network ties, universal rules of reciprocity, shared role making, organizational citizenship behavior, and mental models [1]. HPWS employment is certainly connected with the growth in sales and innovation; conversely, a postulated arbitrating role for employee voluntary turnover is not supported [2]. Lastly, High-Performance Work Systems are deliberated to augment organizational performance by fostering employee proficiency, commitment, and efficiency. The potential of a high-performance work system to assist as an incomparable resource associate with the effective application of corporate strategy and the accomplishment of operative goals [8].

2.2 Technology Startups

New-fangled technology startups are seen as an essential element of the economy [9], technology startups present a fascinating medium, distant from main markets but opulently gifted with human capital. A technology startup can be run on a personal basis in its initial days. The corporate mission and chosen strategy of a technology startup can be successfully carried through direct communication with each employee [10]. An interconnected set of policies focusing on the ensuing problems must be established improving job training, growing employee and union involvement; Growing firms' assurances to stakeholders, building inter-firm collaboration and quality standards and ruling out the low-wage path [11]. Enticing, choosing, and training employees are time and again precarious actions for most technology startups. The study recommends that high-performance work systems enhance organizational performance [4]. Industry 4.0 delivers new standards for the industrial management of technology startups. Sustained by a developing number of innovative technologies, this theory seems more flexible and less costly than traditional enterprise information systems such as ERP and MES [12]. However, technology startups catch themselves ill-equipped to face these new-fangled prospects concerning their production planning and control practices.

The tactical role of HRM, and precisely, the stimulus of an organization's HRM structure on the financial performance, has created significant curiosity within the academic groups [8].

2.3 Objectives of the Implementation of High Performance Works System

Technology startups implement High performance works system with specific goals to achieve, these goals consist (i) Flexibility (ii) Cost reduction (iii) Enhancing productivity (iv) Improving quality (v) Reduce time of delivery. These factors are well documented in the study of Moeuf et al. (2018). The flexibility goal is concentrated towards the ability of the human resource to get adjusted to varied works and situations [13]. The second goal is the reduction of cost through a High-performance works system which has resulted in reducing the cost of manufacturing [14]. Thirdly, improving the productivity in manufacturing through High performance works system [15]. Product quality can be enhanced through training and developing the skills of employees, hence a high-performance works system plays a key role in improving the product quality [16]. Application of a High-performance works system to lean manufacturing improves the production flow and improves delivery [17, 18].

2.4 Industrial Managerial Capacity

Managerial capabilities are significant for the effective employment of technology in technology startups. Alexandre M et al. presented the managerial capabilities structure for understanding Industry 4.0, which also embraces High performance works system, the aspects related to managerial capabilities are (a) monitoring (b) control (c) optimization (d) autonomy. The factor of monitoring in the manufacturing process is achieved through monitoring capabilities. Control as managerial competencies, which offers past data of manufacturing and also indicates the level of acceptance in manufacturing [16, 19]. The subsequent variable of managerial capabilities contains optimization which is dedicated in the direction of refining the process and accomplish a higher level of productivity of manufacturing, outcomes with regards to technology startups have shown an optimistic relationship with regards to productivity and High performance works system. The last, variable related to managerial capabilities is autonomy, which specifies that execution of a High-performance works system entails several techniques and technologies for effective implementation [13, 20–23].

The literature mentioned above on application of High performance works system, objectives of adopting High performance works system and role of managerial capabilities for implementation of High-performance works system, specifies the following questions for the study, which are as under:

Q1: The association amongst application of High performance works system and managerial capabilities in technology startups.
Q2: The role of managerial capabilities in achieving the goals of implementing a High-performance works system in technology startups.

The above discussion through literature review makes this study unique as it shows a combination of the application of High-performance works system to Objectives of High performance works system and role of managerial capabilities in the adoption of High performance works system in technology startups.

3 Research Methods

Based on the literature survey conducted, the following research methodology was adopted to collect and analyze the data.

3.1 Research Design and Instrument

In this research, an exploratory research design is used to comprehend the implementation of a High-performance works system in technology startups. The data were assimilated using a structured questionnaire on a 5 pointer Likert scale.

3.2 Data Collection

The data for the study was collected from 28 technology startups (Break-up is provided in Table 1) that employed a High-performance works system in Karnataka, India.

3.3 Data Analysis Tools

The data collected is studied through a paired sample t-test, which decides the variance between a set of observations. Therefore, in this research variance between the implementation of High performance works system to Goals of application of High-performance works system and managerial capabilities are studied. The use of paired sample t-test is suitable as the data is collected through a Likert scale from technology startups who have implemented a High-performance works system [24]. The summary of the respondents for this research is presented in Table 1. The profile is established on the number of employees working in the organizations under consideration for the present study.

Table 1. Profile of respondents for the study based on "number of employees"

Sl. no.	Profile of the respondents	N	%
1	100–150 employees	11	39.29
2	70–100 employees	7	25.00
3	50–70 employees	4	14.29
4	<50 employees	6	21.42
	Total	28	100

4 Results and Discussions

The results of the study are represented in this segment. Results are shown in Table 2 and Table 3. The results of the analysis are presented concerning the association between the application of the High-performance works system and managerial capabilities essential for effective implementation of HPWS (High-performance works system) and in addition with regards to managerial capabilities and objectives of the implementation of HPWS (High-performance works system).

Operational performance in technology startups with High performance works system and relationship with the managerial capability of Optimization (CAP3) has shown a positive association with Sig. of 0.002 which is less than 0.05 value of significance, though there is an adverse relationship with respect to monitoring where Sig. of 0.129 is greater than 0.05, In the case of Control the Sig. is 0.020 which is less than 0.05 and for Autonomy, the Sig. is 0.025 which is again less than the 0.05 value of significance. The Flow of information has displayed a positive association with monitoring.

The outcomes concerning the flow of information with a High-performance works system and monitoring have presented positive results with a Sig. of 0.00 which is less than 0.05. Conversely, outcomes with respect to control and autonomy are not significantly associated with a value of 0.38 for control and 0.96 for autonomy respectively which is greater than 0.05. However, for the flow of information and optimization, the results have shown a positive association with a Sig. value of 0.004 which is less than

Table 2. Association between the implementation of High performance works system and managerial capability

Groups	Association	Mean	SD	Error	95% confidence interval of the difference		t	DF	Sig. (2-tailed)
					Lower	Upper			
1	HP1 – CAP 1	0.21	1.14	0.13	−0.06	0.47	1.54	72	0.13
2	HP1 – CAP 2	0.43	1.53	0.18	0.07	0.78	2.38	72	0.02
3	HP1 – CAP 3	−0.43	1.11	0.13	−0.68	−0.17	−3.29	72	0.00
4	HP1 – CAP 4	0.37	1.38	0.16	0.05	0.69	2.29	72	0.03
5	HP2 – CAP 1	0.95	1.41	0.17	0.62	1.28	5.72	72	0.00
6	HP2 – CAP 2	0.36	1.44	0.17	0.02	0.69	2.12	72	0.04
7	HP2 – CAP 3	0.58	1.76	0.21	0.17	0.99	2.80	72	0.00
8	HP2 – CAP 4	−0.27	1.39	0.16	−0.60	0.05	−1.69	72	0.10
9	HP3 – CAP 1	0.52	1.69	0.20	0.13	0.92	2.63	72	0.01
10	HP3 – CAP 2	1.10	1.87	0.22	0.66	1.53	5.02	72	0.00
11	HP3 – CAP 3	−1.22	1.60	0.19	−1.59	−0.85	−6.51	72	0.00
12	HP3 – CAP 4	−1.00	1.78	0.21	−1.42	−0.59	−4.80	72	0.00
13	HP4 – CAP 1	−1.85	1.39	0.16	−2.17	−1.53	−11.35	72	0.00
14	HP4 – CAP 2	−1.06	1.86	0.22	−1.49	−0.62	−4.84	72	0.00
15	HP4 – CAP 3	−0.48	1.80	0.21	−0.90	−0.06	−2.28	72	0.03
16	HP4 – CAP 4	0.66	1.63	0.19	0.28	1.04	3.45	72	0.00
17	HP5 – CAP 1	0.88	1.86	0.22	0.44	1.31	4.02	72	0.00
18	HP5 – CAP 2	0.03	1.48	0.17	−0.32	0.37	0.16	72	0.88
19	HP5 – CAP 3	0.82	1.83	0.21	0.40	1.25	3.84	72	0.00
20	HP5 – CAP 4	1.40	1.98	0.23	0.93	1.86	6.02	72	0.00

(Aberrations: HP1: Operational performance, HP2: Flow of information, HP3: Fast evaluation of performance, HP4 Improved decision making, HP5 Integration of work, CAP1 Monitoring, CAP2 Control, CAP3 Optimisation, CAP4 Autonomy, I: Mean, σ: Standard deviation, ϵ: Standard error, DF: Degree of freedom).

0.005. Fast assessment of manufacturing performance has indicated a positive association with respect to control, optimization, and autonomy with Sig. of 0.000 which is less than 0.05. Conversely, with respect to monitoring the results have indicated a negative relationship with a Sig. of 0.10 which is greater than 0.05. Decision making with High performance works system and Managerial capability has presented a positive association with regards to monitoring, control, and autonomy with a Sig. of 0.000 which is less

Table 3. Relationship between managerial capability and objectives of the implementation of high-performance works system

Groups	Association	Mean	SD	Error	95% Confidence interval of the difference		t	DF	Sig. (2-tailed)
					Lower	Upper			
1	CAP1 - OBJ1	−0.38	1.43	0.17	−0.72	−0.05	−2.29	72	0.03
2	CAP1- OBJ2	−0.16	1.29	0.15	−0.47	0.14	−1.09	72	0.28
3	CAP1- OBJ3	−1.01	1.54	0.18	−1.37	−0.65	−5.62	72	0.00
4	CAP1 - OBJ4	−0.22	1.46	0.17	−0.56	0.12	−1.29	72	0.20
5	CAP1 - OBJ5	0.36	1.47	0.17	0.01	0.70	2.08	72	0.04
6	CAP2 - OBJ1	0.27	1.10	0.13	0.02	0.53	2.14	72	0.04
7	CAP2 - OBJ2	0.49	1.49	0.18	0.15	0.84	2.82	72	0.00
8	CAP2 - OBJ3	−0.36	1.15	0.13	−0.62	−0.09	−2.65	72	0.01
9	CAP2 - OBJ4	0.44	1.41	0.17	0.11	0.77	2.65	72	0.01
10	CAP2 - OBJ5	1.01	1.41	0.17	0.69	1.34	6.15	72	0.00
11	CAP3 - OBJ1	0.27	1.11	0.13	0.02	0.53	2.11	72	0.04
12	CAP3- OBJ2	0.49	1.19	0.14	0.22	0.77	3.54	72	0.00
13	CAP3 - OBJ3	−0.36	1.25	0.15	−0.65	−0.06	−2.43	72	0.02
14	CAP3 - OBJ4	0.44	1.21	0.14	0.16	0.72	3.09	72	0.00
15	CAP3 - OBJ5	1.01	1.24	0.15	0.72	1.30	6.98	72	0.00
16	CAP4 - OBJ1	−0.18	1.22	0.14	−0.46	0.11	−1.25	72	0.22
17	CAP4 - OBJ2	0.04	1.55	0.18	−0.32	0.40	0.23	72	0.82
18	CAP4 - OBJ3	−0.81	1.39	0.16	−1.13	−0.48	−4.96	72	0.00
19	CAP4 - OBJ4	−0.01	1.47	0.17	−0.36	0.33	−0.08	72	0.94
20	CAP4 - OBJ5	0.56	1.44	0.17	0.23	0.90	3.33	72	0.00

(Aberrations: CAP1 Monitoring, CAP2 Control, CAP3 Optimisation, CAP4 Autonomy, OBJ1 Flexibility, OBJ2 Cost reduction, OBJ3 improving productivity, OBJ4 Improving quality, OBJ5 delivery time reduction, I: Mean, σ: Standard deviation, ϵ: Standard error, DF: Degree of freedom).

than 0.05. Outcomes concerning optimization and decision making are showing negative results with Sig. of 0.26 which is greater than 0.05. Integration into manufacturing with a High-performance works system and managerial capabilities has shown a positive association with respect to monitoring, optimization, and autonomy with a Sig. of 0.00 > 0.05. But, integration and control of manufacturing activity have shown a negative association with a Sig. of 0.875 which is greater than 0.05.

The results of attaining the goals of employment of a High-performance works system and the role of managerial capabilities are specified in this section. Outcomes with respect to monitoring managerial capabilities and goals of High-performance works system has indicated a positive relationship with respect to improving productivity with Sig. of 0.00 which is less than 0.05 value of significance. Conversely, it shows a negative relationship with respect to flexibility, cost reduction, improving quality, and delivery time with Sig. of 0.25,0.28,0.20 and 0.41 respectively which are greater than 0.05. Managerial control abilities show a positive association concerning cost reduction and delivery time reduction with a Sig. of 0.00 for both which is less than 0.05. The managerial capability of optimization shows a positive association for cost reduction, improving productivity, and reduction in delivery time with Sig. Value of 0.00 which is less than 0.05. Though, a negative relationship is observed concerning flexibility and improved productivity with 0.38 and 0.18 respectively, and which are greater than 0.05. Outcomes with respect to autonomy show a positive association with improving productivity and reduction in delivery time with a Sig. of 0.00 for both and which are less than 0.05 value of significance. On the other hand, it shows a negative association with respect to flexibility, cost reduction and improving quality with a Sig. of 0.215, 0.821 and 0.937 respectively and which are greater than 0.05 value of significance.

4.1 Discussions

The latitude of High-performance works system is further than its application in the technology startups, the areas of research for this study were hooked on to two areas, that is, the relationship between the application of High performance works system and the objective of adopting High performance works system in technology startups and besides, the study was also associated to managerial capabilities essential to implement High performance works system and mapping with goals of application of High performance works system. The outcomes with respect to the implementation of High performance works system and its association with respect to goals of High-performance works system show that the process in technology startups is diligently associated with managerial capabilities i.e., monitoring, control, optimization, autonomy, and the identical is echoed with respect to improved decision making in the process technology startups. Conversely, there are some capacities which necessitate consideration to gain wide-ranging benefits of High-performance works system, which are the integration of human and machines, this would augment operational performance and develop managerial capabilities of technology startups in the attainment of benefits from high performance works system. Consequently, the investigation question with respect to the implementation of a High-performance works system and managerial capabilities show a reasonable level of association and necessitates additional improvement in the flow of information and enhance operational performance.

An additional viewpoint of this research was with respect to managerial capabilities and goals of application of High-performance works system. The results show that managerial capabilities need improvement to attain the goal of the application of High-performance works system technology startups. Despite the fact that the implementation is reasonably effective, this indicates that technology startups have to improve their managerial capabilities for effective application of High-performance works system. Especially with respect to optimization and autonomy.

The alternative role for effective implementation of High performance works system in technology startups is with respect to ascertaining the precise goal for application of High-performance works system. Several firms in the study were not clear with respect to the choice of goals, Most of the technology startups were having multiple objectives, and therefore there was a moderate level of attainment in the application of the High-performance works system.

5 Conclusions

To realize competitive advantage, technology startups must attempt to develop their HRM systems by adopting cutting-edge technology and systems. The current study indicates that technology startups have implemented HPWS (High-performance works system) for augmenting their process. The outcomes of this study have specified that technology startups should understand the association with respect to goals of application of High performance works system in technology startups. The association amongst implementation of High-performance works system and managerial capabilities show that technology startups need to improve managerial skill-sets for effective implementation of High-performance works system. Whereas, the results show that lucidity should be developed with respect to the goals of implementing a High-performance works system in technology startups. The study will further sustain the ineffective implementation of a High-performance works system. Additionally, the study will also make available comprehensive information for augmenting the board area of operations in the added functional area of the technology startups, which will make available a comprehensive understanding of the factors that influence the application of High performance works system in technology startups.

References

1. Evans, W.R., Davis, W.D.: High-performance work systems and organizational performance: the mediating role of internal social structure. J. Manag. **31**(5), 758–775 (2005). https://doi.org/10.1177/0149206305279370
2. Messersmith, J.G., Guthrie, J.P.: High performance work systems in emergent organizations: Implications for firm performance. Hum. Resour. Manage. **49**(2), 241–264 (2010). https://doi.org/10.1002/hrm.20342
3. Ramsay, H., Scholarios, D., Harley, B.: Employees and high-performance work systems: testing inside the Black Box. Br. J. Ind. Relat. **38**(4), 501–531 (2000). https://doi.org/10.1111/1467-8543.00178
4. Bendickson, J., Muldoon, J., Liguori, E., Midgett, C.: High performance work systems: a necessity for startups. J. Small Bus. Strategy **27**(2), 14 (2017)
5. Boxall, P., Macky, K.: Research and theory on high-performance work systems: progressing the high-involvement stream. Hum. Resour. Manag. J. **19**(1), 3–23 (2009). https://doi.org/10.1111/j.1748-8583.2008.00082.x
6. Torre, E.D., Solari, L.: High-performance work systems and the change management process in medium-sized firms. The Int. J. Hum. Res. Manage. **24**(13), 2583–2607 (2013). https://doi.org/10.1080/09585192.2012.744337

7. Qiao, K., Wang, X., Wei, L.-Q.: Determinants of high-performance work systems in small and medium-sized private enterprises in China. Asia Pac. J. Hum. Resour. **53**(2), 185–203 (2015). https://doi.org/10.1111/1744-7941.12038
8. Becker, B.E., Huselid, M.A.: High Performance Work Systems And Firm Performance, p. 25 (1998)
9. Chorev, S., Anderson, A.R.: Success in Israeli high-tech start-ups: critical factors and process. Technovation **26**(2), 162–174 (2006). https://doi.org/10.1016/j.technovation.2005.06.014
10. Davila, A., Foster, G., Jia, N.: Building sustainable high-growth startup companies: management systems as an accelerator. Calif. Manage. Rev. **52**(3), 79–105 (2010). https://doi.org/10.1525/cmr.2010.52.3.79
11. Appelbaum, E., Batt, R.: High Performance Work Systems: American Models of Workplace Transformation. Economic Policy Institute, Washington (1993)
12. Moeuf, A., Pellerin, R., Lamouri, S., Tamayo-Giraldo, S., Barbaray, R.: The industrial management of SMEs in the era of Industry 4.0. Int. J. Prod. Res. **56**(3), 1118–1136 (2018). https://doi.org/10.1080/00207543.2017.1372647
13. Chalal, M., Boucher, X., Marques, G.: Decision support system for servitization of industrial SMEs: a modelling and simulation approach. J. Decis. Syst. **24**(4), 355–382 (2015). https://doi.org/10.1080/12460125.2015.1074836
14. Bonfanti, A., Del Giudice, M., Papa, A.: Italian craft firms between digital manufacturing, open innovation, and servitization. J. Knowl. Econ. **9**(1), 136–149 (2015). https://doi.org/10.1007/s13132-015-0325-9
15. Ren, L., Zhang, L., Tao, F., Zhao, C., Chai, X., Zhao, X.: Cloud manufacturing: from concept to practice. Enterp. Inf. Syst. **9**(2), 186–209 (2015). https://doi.org/10.1080/17517575.2013.839055
16. Segura, D.M., Velandia, N.K., Whittow, W.G., Conway, P.P., West, A.A.: Towards industrial internet of things: crankshaft monitoring, traceability and tracking using RFID. Robot. Comput. Integr. Manuf. **41**, 66–77 (2016). https://doi.org/10.1016/j.rcim.2016.02.004
17. Denkena, B., Dengler, B., Doreth, K., Krull, C., Horton, G.: Interpretation and optimization of material flow via system behavior reconstruction. Prod. Eng. Res. Devel. **8**(5), 659–668 (2014). https://doi.org/10.1007/s11740-014-0545-z
18. Gebauer, H., Fleisch, E.: An investigation of the relationship between behavioral processes, motivation, investments in the service business and service revenue. Ind. Mark. Manag. **36**(3), 337–348 (2007). https://doi.org/10.1016/j.indmarman.2005.09.005
19. MacKerron, G., Kumar, M., Kumar, V., Esain, A.: Supplier replenishment policy using e-Kanban: a framework for successful implementation. Prod. Plan. Control **25**(2), 161–175 (2014). https://doi.org/10.1080/09537287.2013.782950
20. Barenji, A.V., Barenji, R.V., Roudi, D., Hashemipour, M.: A dynamic multi-agent-based scheduling approach for SMEs. Int. J. Adv. Manuf. Technol. **89**(9–12), 3123–3137 (2016). https://doi.org/10.1007/s00170-016-9299-4
21. Behbahani, S., de Silva, C.W.: System-based and concurrent design of a smart mechatronic system using the concept of mechatronic design quotient (MDQ). IEEEASME Trans. Mechatron. **13**(1), 14–21 (2008). https://doi.org/10.1109/TMECH.2007.915058
22. Brad, S., Murar, M.: Employing smart units and servitization towards reconfigurability of manufacturing processes. Procedia CIRP **30**, 498–503 (2015). https://doi.org/10.1016/j.procir.2015.02.154
23. Bughin, J., Chui, M., Manyika, J.: An executive's guide to the Internet of Things. McKinsey Q. **4**, 92–101 (2015)
24. Roberson, P.K., Shema, S.J., Mundfrom, D.J., Holmes, T.M.: Analysis of paired Likert data: how to evaluate change and preference questions. Fam. Med. **27**(10), 671–675 (1995)

Optimization of Support Vector Machine for Classification of Spyware Using Symbiotic Organism Search for Features Selection

Noah Ndakotsu Gana[1], Shafi'i Muhammad Abdulhamid[1], Sanjay Misra[2](✉), Lalit Garg[3], Foluso Ayeni[4], and Ambrose Azeta[2]

[1] Federal University of Technology, Minna, Nigeria
noah.gana@st.futminna.edu.ng, shafii.abdulhamid@futminna.edu.ng
[2] Center of ICT/ICE Research, Covenant University, Ota, Nigeria
{sanjay.misra,ambrose.azeta}@covenantuniversity.edu.ng
[3] University of Malta, Msida, Malta
lalit.garg@um.edu.mt
[4] University of Nebraska, Omaha, USA
fayeni@unomaha.edu

Abstract. Malware's key target is to compromise system security pillars, the confidentiality, integrity and availability. Spyware is a form of malware program that collect entity's information including personal confidential information, activity logs on computing system, financial transaction, password and geolocation precision through monitoring target without prior knowledge of victim. The integration of computing devices into daily existence, as well as the exponential development experienced in application development including the expansion of interconnected computing devices serve as goldmine to malicious entities for target and exploit using spyware. In previous literature, Support Vector Machine (SVM) was employed for the classification of spyware, but has suffered setbacks of low performance as a result of untuned parameters as well as the use of irrelevant dataset features for training and classification. The optimization of SVM for classification of spyware using Symbiotic Organisms Search (SOS) algorithm for feature selection was therefore deployed to enhance performance. The results obtained from the study indicate that the technique performed optimally in spyware classification recording the following; 97.40% and 2.3% respectively for accuracy and false positive rate respectively. Therefore, revealed that the optimization of SVM with SOS for classification enhances performance and reduces the rate of false alarm which is an improvement on existing literatures. This points the fact that tuned parameters of the model can be implemented for proper classification of spyware attacks.

Keywords: SVM · SOS · Spyware classification · Accuracy · False positive rate

1 Introduction

Sustainability Compromise of system Confidentiality, Integrity and Availability (CIA) is normally the key target of malware according to [1]. A malware can be referred to as

L. Garg et al. (Eds.): ISMS 2020, LNNS 303, pp. 11–21, 2022.
https://doi.org/10.1007/978-3-030-86223-7_2

malicious software that execute series of codes in systems that have been compromised by its nefarious activities, exploiting the security defense mechanism in place. Virus, worm, trojan, rootkit and ransom-ware according to [2, 27], are examples of malware for which this research focuses mainly on spyware.

Spyware is a program that collects entity's information ranging from personal information, activities performed by entity, financial transaction, password, geo-location precision through monitoring without entity's prior knowledge [3]. Spyware was first recorded in 1995 by Microsoft's business model which denotes as an espionage software, also spyware is an espionage ransomware code that exfiltrates sensitive information.

Spyware software are akin to malicious program as it entices users into application execution, while being stealthy by circumventing removal activities as it uses subliminal channel. Although some spyware program is embedded in End User License Agreement (EULA) as backdoors to obtain user's consent, some spyware is installed without the user's consent. Spyware also have the capability of transmitting harvested information to third party after stealthily monitoring user behavior, web surfing habits, confidential details and user profile.

The exponential engagement of the world population in Information Communication and Technology (ICT) usage as at the 2018 was 51.2% which is 3.9 billion population, there have been a steady upward trend in the incorporation and usage of ICT, likewise cybercrime projection is estimated at USD 2 billion by the end of 2019, as a result of skyrocketing rate of ICT usage [4]. This becomes a challenge to ICT users based on the fact that malicious activities need to be exterminated to avoid breach of CIA, spyware which is one of the stealthy unwitting trending malicious activity needs to be checkmated.

Support Vector Machine (SVM) a supervised machine learning model and noted for a, map feature vectors from nonlinear space to a higher multidimensional dimensional space, thereby making use of linear classifier obtained from the new space, suppose H represents the generated new feature space and $\varnothing$ represents mapping function so that $\varnothing : R^d \rightarrow H$, the feature vector $\overline{x} \in R^d$, the mapping of feature vector is denoted by $\varnothing(\overline{x})$, while the y label stands same, thus, $(\overline{x}_i, y)$ which is the training sample becomes$(\varnothing(\overline{x}_i), y)$, furthermore, H defines the hyperplane in the transformed space, which segregates the training sample $(\varnothing(\overline{x}_i), y_i), \ldots, (\varnothing(\overline{x}_n), y_n)$. This leads to obtaining a hyperplane in the space H which permits the mapping of feature vector $\varnothing(\overline{x}_i)$ to be segregated on one side of the hyperplane for label $y_i = -1$, and $\varnothing(\overline{x}_i)$ to the other side of the hyperplane for $y_i = 1$, [5, 6].

However, SVM perform better once the parameters such as the kernel function are optimized and optimal feature for classification are also well defined [7–9, 25].

Symbiotic Organism Search (SOS) is a metaheuristic algorithm that is based on organism symbiotic association in an ecosystem widely employed in various fields for optimization of problems ranging from scheduling of task, construction project and engineering structure design optimization. SOS was first introduced by [10, 11] and refer to as a symbiotic based relation of organisms in ecosystem for numerical optimization and engineering design problem, an initial population known as the ecosystem is defined at the initialization stage, which further generates a random organism population, for each related problem, an organism stands as a candidate solution which indicates the adaptiveness degree of an organism, each organism is bound to pass through the three major phases of symbiotic interaction namely; mutualism, commensalism and parasitism respectively in an iterative process, symbiotic organism search algorithm is effective solution to solve complex numeric computations regardless of it few control parameters compared to some other optimization algorithms [12].

Grid search algorithm to an exhaustive search algorithm, which completely explores a search space, while the variable to be optimized is represented by each dimension of a grid coordinate, the grid search works with a define range of value, known as the maximum and minimum value, that aids in establishing an optimal variable [13].

Grid search algorithms concept is based on setting the parameter values such as the SVM kernel function of C, γ and step sizing, in order to determine a grid search points. Thus, for each parameters (C, γ) in the grid Support Vector Machine model is trained, in which the sample data is evaluated using the optimal selected model of training results [14, 26]. Therefore, this research intends to apply SOS for selection of optimal spyware features that will be trained for classification.

Summary of the key contribution of this study are outlined below;

- We design SOS metaheuristic algorithm for spyware feature selection which obtain optimal features
- We Optimized SVM classifier in order to achieve a superior classification performance over default SVM parameter.

United the remaining sections of the paper were organized as follow Section 2 presents related literature in spyware classification, Sect. 3 reveals the methods employed in the research, Sect. 4 presents the results obtained in performance evaluation of spyware classification and Sect. 5 shows the conclusion and recommendation.

2 Related Literature

[3] in an experiment to classify spyware affected files through the implementation of data mining technique, more than eight thousand malware samples with hundreds of benign sample were used in spyware classification, which was based on Application Programming Interface (API) call dataset, J48 Decision Tree classification algorithm was used, an accuracy, true positive rate and false positive rate of 86.93%, 86.9% and 3.3% respectively was achieved, however, there exist an imbalance dataset sample based on malware to benign ratio, while focus was majorly based on API system call.

[14] proposed a SpyAware framework that encompasses of a profiler, a feature extraction and a classifier, which were to aid in automatic profiling of app execution as it relates to binder calls and system, calls, obtaining feature vectors from execution traces and predicting and training of spyware execution in terms of feature vectors; support vector machine (SVM) and Naïve Bayes classifiers was incorporated at the classification stage. Furthermore, performance level of 67.4% and 64.2% accuracy was achieved in detecting spyware execution respectively, however, the research focuses majorly on smartphone privacy leakage issues, and also based on a define version of Android OS platform, accuracy rate is low and a high undisclosed FPR is said to be achieved.

In order, to counteract the challenges faced by existing anti-spyware tools such as detecting of spyware that have capacity to modify self against detection, [15] proposed a Stateful Threat Aware Removal System (STARS) that has the potent to track critical activities of running process, monitoring spyware removal task effectiveness over a period, and establishing a trade-off between system dependability and system performance as it relates to severity threat of spyware system. To this end, it was establish that STARS has the capacity to detect and remove self-healing spyware a challenge faced by existing commercial anti-spyware tools based on an experiment performed in this study, however, there is a shortfall in overall performance in removing self-healing spyware likewise the proposed method lack the capacity to detect hidden registry entries.

[16] opine a prevention mechanism against key logger spyware attacks, the proposed methodology include the following phase, key logger spyware attack, honeypot based detection and prevention of key logger spyware, generation of spyware attack in order to help track attack behavior, detection of keylogger spyware, monitoring of malicious system activities and permanently disabling the key logger spyware by a prevention server is achieve respectively by the aforementioned phases, it was stated that proposed mechanism if employ can tackle key logger spyware attack, however, focus was designated on key logger spyware attack alone.

Cloud theory model was used to develop an interest model to enable spyware detection, the research, [17] present a novel spyware detection technique that employs an abstract characterization of popular classes of spyware programs through the use of data mining approach as a result of its capacity to discover program of interest in large amount of behaviors, thereby leading to overcoming the drawback associated with unknown signature based detection as the proposed model can detect unknown spyware as well as variant of known spyware theoretically, it was further reveal that the define model was able to detect spyware programs optimally, though, this research was theoretically based and not implemented in real system scenario.

Extraction and selection of optimal features to detect spyware was proposed by [18] in the research that leads to optimal features selection is based on the frequency and appearance of the feature in the dataset as opine in this study. Accuracy performance metrics was employed in this study as well as the following classification algorithms; ZeroR, Naïve Bayes, C4.5 decision Tree (J48), Support Vector Machine (SVM), JRip and Random Forest attaining an accuracy of 91.50%, 99.49%, 99.86%, 99.80%, 99.24% and 99.86% respectively with n-gram equal to 5 out of 100 selected features, J48 classifier outperforming all competing approaches, however, more performance metrics will enhance result interpretation and gauging,

In order to attain an optimal and accurate detection of Adware, data mining algorithms such as Naïve Bayes, Support Vector Machine algorithm SMO, IBk, J48, and JRip were employed in the proposed approach for accurate detection of Adware using Opcode sequence extraction to identify unseen and novel instances of adware along n-gram size, detection rate, false alarm rate, and accuracy were used as performance evaluation metric including area under receiver operation characteristics curve (AUC). ZeroR serve as the baseline classifier, IBk achieve AUC, FNR, FAR of 0.949, 0.022 and 0.115 respectively with $n = 4$ and a 70% split. IBk was said to have outperform other classification algorithms used, however, focus was based on adware, with a little collection of dataset used in experimental evaluation [19].

The research performed by [19] basically based on evaluation of Windows platform executable achieve through the implementation of machine learning algorithms such as ZeroR, Naïve Bayes, Support Vector Machine (SMO), J48, Random Forest, and JRip classification algorithms, also, a 10 fold cross validation was employ to classify unseen binaries. Accuracy and Area Under Receiver Operation Characteristics (ROC) curve was used as metrics in terms of performance evaluation. J48 classification algorithms achieve 90.5% accuracy using n as 6, denoting the highest accuracy compared to other classification algorithms used in this study, while ZeroR, Naïve Bayes, SVM, Random Forest and JRip achieved 86.92%, 89.80%, 89.65%, 89.48%, 89.45% accuracy respectively, Random Forest algorithm give an AUC score of 0.83 using n as 6,ZeroR, Naïve Bayes, SVM, J48, Random Forest and JRip achieved AUC of 0.50, 0.62, 0.71, 0.65 and 0.66 respectively furthermore, common feature based extraction and frequency feature extraction was employ in order to obtain Reduced Feature Set (RTS) which was further used in generation of arff files, nevertheless, attention dwell mostly on Windows executable and dataset used in experiment is of small size.

[20] in their research opined a framework with the capability of detection and classification of spyware, the following classification algorithms Decision Tree, ZeroR, JRip, J48 and Naïve Bayes was applied in classifying existing spyware, Decision Tree attained the best accuracy of 97.7854, Kappa Statistic of 0.723 and ROC area of 0.9356, which serve as a robust rule based algorithm to enhance the proposed framework.

Data mining based detector optimized by Breadth-First Search algorithms was employed to achieve an accuracy of 90.5% and 0.731 FPR in the research by [21] in order to detect spyware, feature set generate form Common Feature-based Extraction (CFBE) feature selection technique with n = 4 achieve the accuracy of 89.49%, 88.21% and 88.02% respectively for Random Forest, Naïve Bayes, and Support Vector Machine, the following FPR was also recorded against each of aforementioned classifier respectively 0.731, 0.665 and 0.665 was used as a comparative factor, however, the research experiment was majorly based on executable files in evaluating the performance of employed method that was developed and experience a high FPR and low accuracy. [1] proposed a kernel level system routine interception in detecting and eliminating spyware and ransom ware, Linear Regression, JRip and J48 decision tree classifiers was employed in the research in order to achieve the spyware and ransom ware detection as well as elimination, experiment performed based on the designed methodology give an accuracy of 93% with a FPR of 7%, however, the resulting performance evaluation indicates a low accuracy couple with high FPR.

[22] opined a surveillance spyware detection system that encompasses both static and dynamic analysis, in order to classify spyware SVM classification algorithm optimal features generated from information gain ranking was trained, an accuracy of 97.91% for known spyware and 96.4% for the unknown spyware was achieved and a false positive rate of 0.68% and TPR of 95.33% based on the static and dynamic analysis performed, however, the research based the experiment basically on executable and a resulting high FPR. [23] proposed a hypothesis of the possibility of classifying software that have spyware functionality embedded based on the software End User License Agreement

(EULA), the experiment performed by the study using data information obtained from 100 software application through the means of anti-spyware application in order to aid in determining software applications that have spyware features embedded therein based on the application EULA, 15 different classification algorithms was employed and multi-nominal Naïve Bayes, SVM and Voter Perception algorithms was assert to have outperformed other classification algorithms while achieving optimal AUC and accuracy rate and low false positive rate, the obtained result indicates that the proposed hypothesis about EULAs can serve as an evaluating mechanism for classification of software with embedded spyware functionality, however the result indicates a very high FPR and Low accuracy.

[24] developed a novel malware detection technique known as Memory management with API Call mining (MACA-I) to detect malware that transit in memory management API, monitored and tracked based on dynamic analysis, the evaluation of the developed technique was based on accuracy and sensitivity performance metrics, using the following machine learning algorithms; Logistics Regression, Support Vector Machine (SVM), and Decision Tree, which achieved the following accuracy 78.78%, 77.27% and 89.89% respectively, while the sensitivity of 91.17%, 85.28% and 97.05% was attained respectively, however, the research majored on API calls only. Other works exist in which SVM was applied. Such works include [28] in which SVM for applied for the predict6ion of path loss while the work of [29] applied a variant of SVM called support vector regression (SVR) for the forecasting of stream flow.

3 Methodology

In this research, publicly available dataset of Advance Persistent Threat (APT) dataset accessible for research and development found in Microsoft subsidiary repository known as Github, the dataset is comprised of 189 attributes and 291 instances, the following form part of the attributes; techniques, tactics and procedure (TTP), spyware, backdoor, Trojan and rootkit.

Symbiotic Organism Search algorithm and Grid Search algorithm was employed for optimal feature selection and optimal parameter respectively in order to achieve optimization in classification based on SVM. A 5 fold and 10 fold cross validation was employed in this study, a total of 186 optimal features was selected out of 189 feature based on SOS algorithms through the iterative process of 50 and a population size of 50.

In order to optimize learning the dataset was preprocessed, Synthetic Minority Oversampling Technique (SMOTE) was employed to addressing biasness associated with imbalanced dataset, likewise the following metrics was used to remarkably evaluate the performance optimization of SVM for classification of spyware; accuracy, true positive rate and false positive rate.

4 Results

The Tables 1 and 2 below indicates the optimized parameters of both gamma (γ) and cost (c) based on defined range as well as the values of cross validation and the default SVM parameter (γ) and (c) respectively.

Table 1. Optimized SVM parameters Gamma (γ) and cost (c)

Cross Validation (CV)	Gamma (γ)	Cost (C)	Boundary
5	0.1	1000	$1e^{-05} - 1e^{05}$
10	0.1	1000	$1e^{-05} - 1e^{05}$

Table 2. Default SVN parameters (γ) and cost (c)

Gamma (γ)	Cost (C)
0.0	1

4.1 Performance of Spyware Classification Based on Default SVM Parameter

The default SVM parameter was used in spyware classification based on the balance spyware dataset without feature selection and the balanced optimal feature spyware dataset obtained from SOS. The Fig. 1 below depicts the obtained result in this study.

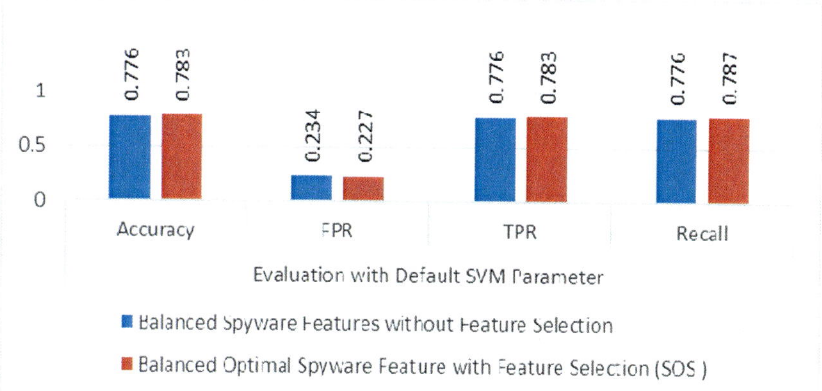

Fig. 1. Spyware classification (balanced spyware features without feature selection and balanced optimal spyware features with feature selection) with default SVM Parameter

4.2 Performance of Spyware Classification Based on Optimized SVM Parameter

The Fig. 2, depicts the performance evaluation result obtained from the classification of spyware through the deployment of the optimized SVM parameters.

To further establish the strength of the optimized SVM parameter for spyware classification in terms of the enhanced performance the Table 3 below represents a performance comparison against existing spyware model classifiers with regards to some relevant performance metrics found in literatures.

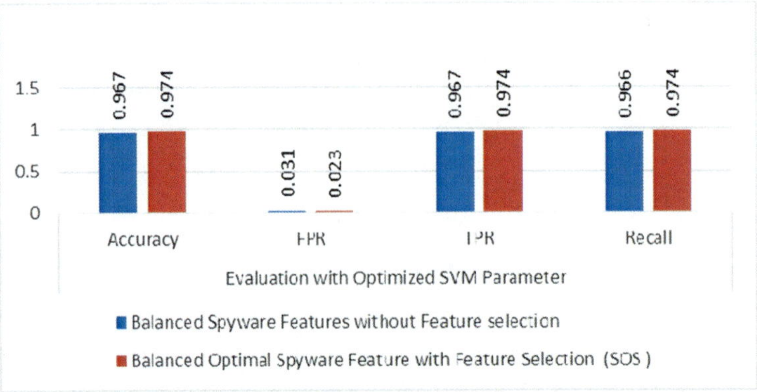

Fig. 2. Spyware classification (balanced spyware features without feature selection and balanced optimal spyware features with feature selection) with default SVM parameter

Table 3. Comparison of proposed optimize SVM classifier with baseline literatures

Reference	Approach	Accuracy	TPR	FPR	Sensitivity or recall
Proposed Optimized Model	**GridSearchSVM Model**	**97.40**	**97.40**	**2.30**	**97.40**
Kumar et al. (2019)	J48 Decision Tree	86.93	86.69	3.30	-
Sai et al. (2019)	MACI-I	95.45	-	-	-
	Logistics Regression	78.78	-	-	91.17
	SVM	77.27	-	-	85.28
	Decision Tree	89.89	-	-	97.05
Javaheri et al. (2018)	Linear Regression + JRip + J48	93.00	-	7.00	92.77

N.B: (-) means the metric value is not reported in the reference.

5 Conclusion and Recommendation

Spyware is a wide spreading stealthy threat to computing environment and devices, the large collect of data from the computing devices may be a challenging factor in terms of redundancy and unbalances of data, which in turn have a great impact and setback to performance analysis evaluation in terms of classification of spyware by machine learning model, not ruling out the tuning of classification model for better performance, hence, this research proposed an enhanced SVM classifier for spyware classification based on Gridsearch optimization algorithm, tuning and determining the optimal SVM parameter, further supported by optimal feature selection based on SOS algorithm in order to enhance spyware classification with a better accuracy and low FPR which is a setback in existing classification model for spyware classification.To achieve an optimal SVM model for spyware classification a Gridsearch optimization based algorithm was implemented in identifying optimal parameters based on defined range of computation for the parameters gamma (γ) and cost (C) function respectively for SVM, while a cross validation was deployed for validate the trained spyware features while obtaining optimal accuracy which defines the optimal SVM parameters of gamma (γ) and cost (C), furthermore, SOS based metaheuristic algorithm selected optimal spyware features used for classification. The recorded research proves that optimal SVM classification algorithm for spyware classification outperformed the traditional SVM classifier and most of the baseline techniques for spyware classification. The supremacy of the performance of the proposed optimal SVM classification algorithm for spyware classification is based on the fact that parameters was tuned, spyware dataset was balanced based on SMOTE, and optimal spyware features was obtained based on SOS metaheuristic algorithm which climax the performance of SVM classification accuracy.

Based on the finding of this research, the research establishes that optimal SVM classification algorithm for spyware classification is an effective model for spyware classification while feature selection-based SOS metaheuristic algorithm serves as better algorithm aids in mitigating the challenges of data overfitting by dropping redundant spyware features. Furthermore in future study, other optimization algorithms alongside couple with other feature selection algorithm can be explored in other to achieve a higher performance in terms of spyware classification.

Acknowledgement. The authors appreciate the sponsorship from Covenant University through its Centre for Research, Innovation and Discovery, Covenant University, Ota Nigeria.

References

1. Javaheri, D., Hosseinzadeh, M., Rahmani, A.M.: Detection and elimination of spyware and ransomware by intercepting Kernel-level system routines. IEEE Access **6**, 78321–78332 (2018)
2. Nawfal, T.O., Wesam, B.: Review of data mining techniques for malicious detection. Res. J. Appl. Sci. **11**(10), 942–947 (2016)
3. Kumar, D.A., Kapat, S.K., Das, S.K., Tripathy, S.N.: Classification of spyware affected files using data mining techniques. Int. J. Recent Technol. Eng. **8**(2), 462–466 (2019). https://doi.org/10.35940/ijrte.B1088.0782S619
4. Guha, D., Roy, P.K., Banerjee, S.: Symbiotic organism search algorithm applied to load frequency control of multi-area power system. Energy Syst. **9**(2), 439–468 (2017). https://doi.org/10.1007/s12667-017-0232-1
5. Arya, M., Bedi, C.S.S.: Survey on SVM and their application in image classification. Int. J. Inf. Technol., 1–11 (2018).https://doi.org/10.1007/s41870-017-0080-1
6. Kulkarni, S.R., Harman, G.: Statistical learning theory: a tutorial. Wiley Interdisc. Rev. Comput. Stat. **3**(6), 543–556 (2011)
7. Alwan, H.B., Ku-Mahamud, K.R.: Integrated ACOR/IACOMV-R-SVM algorithm. Int. J. Comput. Inf. Eng. **11**(12), 1309–1313 (2017)
8. Huang, C.L., Wang, C.J.: A GA-based feature selection and parameters optimization for support vector machines. Expert Syst. Appl. **31**(2), 231–240 (2006)
9. Lin, J., Zhang, J.: A fast parameters selection method of support vector machine based on coarse grid search and pattern search. In: 2013 4th Global Congress on Intelligent Systems, pp. 77–81. IEEE (December 2013)

10. Boswell, D.: Introduction to Support Vector machines. Department of Computer Science and Engineering, University of California, San Diego (2002)
11. Cheng, M.Y., Prayogo, D.: Symbiotic organisms search: a new metaheuristic optimization algorithm. Comput. Struct. **139**, 98–112 (2014)
12. Liu, R., Liu, E., Yang, J., Li, M., Wang, F.: Optimizing the hyper-parameters for SVM by combining evolution strategies with a grid search. In: Huang, D.-S., Li, K., Irwin, G.W. (eds.) Intelligent Control and Automation, pp. 712–721. Springer, Heidelberg (2006). https://doi.org/10.1007/978-3-540-37256-1_87
13. Yuanyuan, S., Yongming, W., Lili, G., Zhongsong, M., Shan, J.: The comparison of optimizing SVM by GA and grid search. In: 2017 13th IEEE International Conference on Electronic Measurement & Instruments (ICEMI), pp. 354–360. IEEE (October 2017)
14. Xu, H., Zhou, Y., Gao, C., Kang, Y., Lyu, M.R.: SpyAware: investigating the privacy leakage signatures in app execution traces. In: 2015 IEEE 26th International Symposium on Software Reliability Engineering (ISSRE), pp. 348–358. IEEE (November 2015)
15. Wu, M.W., Wang, Y.M., Kuo, S.Y., Huang, Y.: Self-healing spyware: detection, and remediation. IEEE Trans. Reliab. **56**(4), 588–596 (2007)
16. Wazid, M., Sharma, R., Avita Katal, R.H., Goudar, P.B., Tyagi, A.: Implementation and embellishment of prevention of Keylogger spyware attacks. In: Thampi, S.M., Atrey, P.K., Fan, C.-I., Perez, G.M. (eds.) Security in Computing and Communications, pp. 262–271. Springer, Heidelberg (2013). https://doi.org/10.1007/978-3-642-40576-1_26
17. Wang, X., Chen, J.: Interests-based spyware detection. In: 2009 International Forum on Computer Science-Technology and Applications, vol. 2, pp. 175–178. IEEE (December 2009)
18. Sheta, M.A., Zaki, M., El Hadad, K.A.E.S.: Spyware detection by extracting and selecting features in executable files. In: 10th International Conference on Electrical Engineering, ICEENG 2016, vol. 10, pp. 1–20. Military Technical College (April 2016)
19. Shahzad, R.K., Haider, S.I., Lavesson, N.: Detection of spyware by mining executable files. In: 2010 International Conference on Availability, Reliability and Security, pp. 295–302. IEEE (February 2010)
20. Tripathy, S., Kapat, S., Das, S., Panda, B.: A Spyware detection system with a comparative study of spywares using classification rule mining. Int. J. Sci. Eng. Res. **7**, 179–184 (2016)
21. Bahraminikoo, P., Samiei, M., Babu, G.P.: Utilization data mining to detect spyware. IOSR J. Comput. Eng. (IOSRJCE) **4**(3), 1–4 (2012)
22. Wang, T.Y., Horng, S.J., Su, M.Y., Wu, C.H., Wang, P.C., Su, W.Z.: A surveillance spyware detection system based on data mining methods. In: 2006 IEEE International Conference on Evolutionary Computation, pp. 3236–3241. IEEE (July 2006)
23. Boldt, M., Jacobsson, A., Lavesson, N., Davidsson, P.: Automated spyware detection using end user license agreements. In: 2008 International Conference on Information Security and Assurance, ISA 2008, pp. 445–452. IEEE (April 2008)
24. Sai, K.N., Thanudas, B., Sreelal, S., Chakraborty, A., Manoj, B.S.: MACA-I: a malware detection technique using memory management API call mining. In: 2019 IEEE Region 10 Conference (TENCON), TENCON 2019, pp. 527–532. IEEE (October 2019)
25. Maigida, A.M., Olalere, M., Ismaila, I.: An intelligent crypto-locker ransomware detection technique using support vector machine classification and grey wolf optimization algorithms. iManager's J. Softw. Eng. **13**(3), 15–23 (2019)
26. Syarif, I., Prugel-Bennett, A., Wills, G.: SVM parameter optimization using grid search and genetic algorithm to improve classification performance. Telkomnika **14**(4), 1502 (2016)
27. Takeuchi, Y., Sakai, K., Fukumoto, S.: Detecting ransomware using support vector machines. In: Proceedings of the 47th International Conference on Parallel Processing Companion, pp. 1–6 (August 2018)

28. Abolade, R.O., Famakinde, S.O., Popoola, S.I., Oseni, O.F., Atayero, A.A., Misra, S.: Support vector machine for path loss predictions in urban environment. In: Gervasi, O., et al. (eds.) Computational Science and Its Applications – ICCSA 2020: 20th International Conference, Cagliari, Italy, July 1–4, 2020, Proceedings, Part VII, pp. 995–1006. Springer International Publishing, Cham (2020). https://doi.org/10.1007/978-3-030-58820-5_71
29. Mohd Radzi, M.R.B., Shamshirband, S., Aghabozorgi, S., Misra, S., Akib, S., Kiah, L.M.: Potential of support-vector regression for forecasting stream flow. Tehnicki vjesnik/Technical Gazette **21**(5) (2014)

Intelli-Helmet: An Early Prototype of a Stress Monitoring System for Military Operations

Akib Zaman(✉), Rafat Tanjim Khan, Nazmul Karim, Muhammad Nazrul Islam, Md Shihab Uddin, and Md Mehedi Hasan

Department of Computer Science and Engineering, Military Institute of Science and Technology, Mirpur Cantonment, Dhaka, Bangladesh

Abstract. With the development of sophisticated technology, battle field is becoming more complex day by day and soldiers' stress management has become an inevitable part of combat efficiency. Real time operation analysis of stress factors is a step towards future soldier system and can give a upper hand for making precise decision by the Commanding Officer during a military operation. Therefore, the objective of this research is to develop a early prototype of stress monitoring system named Inteli-Helmet for soldiers deployed in a military operation. The proposed system incorporates the brain waves and physiological data of (deployed) soldiers to measure their stress and facilitate the higher commander to take the best decision about a soldier's deployment (or return back) in (from) war field. The Electroencephalography (EEG) based Brain-Computer Interfacing (BCI) technology was used to acquire the brain signals, while machine learning was adopted to facilitate the decision making process for the higher commander. A light weight evaluation study was also carried out where it has been found that, the identified stress factors and the decision assistance procedures of the proposed system were effective and efficient for taking decision during military operation by the higher command. Again, the study found that the Bayes Net algorithm showed the higher accuracy of 97.4% followed by J48 Decision Tree algorithm of 96.1% while classifying the soldiers' stress status.

Keywords: Combat efficiency · Future soldier system · Stress monitoring · Brain waves · Electroencephalography · Brain-computer interfacing · Machine learning

1 Introduction

Stress and trauma is very conjoint in the military profession due to involvement of several intensive operations and training programs. Exponential increment of the battlefields related to Fighting in Build-up Area (FIBUA), unconventional warfare, dreadful capability of Improvised Explosive Device (IED), pervasive use

L. Garg et al. (Eds.): ISMS 2020, LNNS 303, pp. 22–32, 2022.
https://doi.org/10.1007/978-3-030-86223-7_3

of snipers in wider spectrum of battle environment can make it extremely hard for a higher commander to appreciate the mental fitness of a soldier for maximizing the combat efficiency and thereby ensuring necessary protection of under commands (UC). It can be also very difficult to ensure maximum military training during peacetime as there is no certain data (pattern) analysis for the evaluation of stress limit during any training procedure. Thus for ensuring upgraded training facilities and to be more adaptive in the stressed situation, study of the stress level of soldiers during various Operations can bring up ground-breaking results and can open up a new window for the military research.

Brain-computer interface (BCI) is a collaboration between a brain and a device that enables signals from the brain to direct some external activity. The EEG (electroencephalography), MEG (magneto encephalography), or MRT (magnetic resonance tomography), etc. are primarily used to develop a noninvasive brain communication interface, while the EEG band technology is used mostly in dumpling BCI band applications [8]. For example, BCI tool which uses the user experience of playing several games, simultaneously has been used in robotic arms to assist individuals who have lost ability to perform everyday activities such as walking because of extreme medical injuries [9]. Likewise, BCI has been extensively used in different aspects of Military enhancements. Very recently, US military emerging technology agency named Defense Advanced Research Projects Agency (DARPA) is investing tens of millions of dollars in developing a brain-computer interface for use by soldiers [10].

In [1], O'Neil et al. conducted a review study and highlighted the key factor of Traumatic Brain Injury (TBI) of the military personnel and veterans; while in another work [2], Bryan and Clemans explored the pattern of suicide attempt taken by the deployed military personnel after completing the deployment. However, all these were carried out in a post-operative environment where the damage has been already endured. Thus further research is required focusing on the operative environment analysis to monitor stress factors and detect stress while prevailing on ground (for a military operation or training) and thereby impact the decision making process.

Therefore, the objectives of this research are to explore the stress factors using brain waves and to develop a soldier's stress monitoring system using brain waves and physiological signals. The proposed system named 'Intelli-Helmet' will be used during a military operation or any extensive military exercise by the higher command to take appropriate decision about a soldier's initial deployment or the continuation of deployment. The research is organized as follows. Firstly, the related works for this research is presented in Sect. 2. Next, the requirement elicitation of the proposed system has been described in Sect. 3. After that, design and development of the system has been discussed in Sect. 4. Finally, the main findings, implications, limitations and the possibility for future research are discussed in Sect. 5.

2 Related Works

This section briefly discuss on few of the prevailing works on BCI on emotions and stress detection both in non-military and military environment with a view to exploring possible research scopes and validation of the development motivation of proposed system. Previous works conducted using BCI are mainly finding out the emotional condition in various context based on brain waves, heart rate and other physiological factors.

Preeti et al. in [7] has recognized the emotion based on EEG signals and applied Random Forest algorithm to classify four classes of emotions i.e. happy, sad, exciting and hate states. Again, Murugappan et al. [11] has worked on identifying different human emotions using EEG using discrete wavelet transform. They used 3 of the brain waves namely Alpha, Beta, and Gamma wavelets using an audio-visual system and could find 05 of human emotions which were happy, disgust, surprise, fear and neutral. The maximum subsets of classification rate were 91.67% for disgust, 81.67% for happy and surprise, 81.25% for fear and 93.75% for neutral. Diya and Prorna in [5] developed a system based on BCI to detect emotions while undergoing user Experience (UX) evaluation in a serious game. Point to be noted that, these works were based on non-military environment which is not as rigorous in nature like the military operations or training exercise.

Bos and Danny et al. in [13] worked with EEG based generated data to identify emotions while playing music to found out the different types of music affecting mental condition of subjects concurrently with Heart rate variability(HRV). On the other hand, Sriramprakash.S et al. in [12], conducted research on working people using GSR, Heart rate and its variability features to find immediate response for stress prediction. Choi and Kim et al. In [3] has tried to explore how accurately human emotions can be measured basing only on physiological data such as HRV. These works have shown that there is a significant relation between stress management and physiological data like HRV, Heartrate etc. However, they have not explored the significance of that relation in a military operational environment.

However, Implications of BCI in Military Operations is pertinent through previous years. There are several articles where BCI has been used for the advancement of military. Razzak and Islam in [4] developed a communication system with visual feedback from battlefield to monitor the soldiers deployed in operations and provide them with real-time guidelines from command base. Islam et al. in [6] developed a sensor-based system for commander to deploy, locate and monitor the soldiers and to aware the status (dead or alive) of soldiers during a military operation. However, these studies have focused on monitoring and gathering operational information rather than individual soldier's mental health analysis in an ongoing operational environment.

In Summary, recent studies have shown that BCI and physiological data like HRV, Heart Rate has been found efficient to detect emotional and stress status. However, there are scopes to explore this significance in case of military operations especially in monitoring mental health of soldiers. It can also be understood that real-time stress monitoring system in a military operation has not been developed yet which could be a life-saving tool for thousands of soldiers. Therefore, this research has been focused to explore the necessary features of this desired system and has developed a BCI and heart rate based system to predict stress status of deployed soldiers in military operation.

3 Requirement Elicitation

The purpose of the requirement elicitation study was to explore the scope and situations of the field and to reveal the features that needs to be incorporated for developing a system to monitor the stress of soldiers operating in the war field.

We interviewed 15 officers as test participants who are currently holding the ranks of Lieutenant Colonel and Major in military for the interview who are between the age group of 30 to 45. All of them were male. They have served in at least one UN missions with active military operation experience. Among them there were four military doctors with service length of 12 to 15 years who are specialized in different arenas of Medical Science including Medicine and Neurosurgery. They have experience of using electronic gadgets in different spheres of their life. The interview session was taken one by one. The sole intention of the interview was to bring out real life requirements of the proposed system and personal privacy was maintained strictly. Furthermore, 38 officers and 20 health personnel were given with a structured closed ended Google form to response. It was ensured that misuse of any response will be prevented and will be only used for research purpose.

This requirements elicitation study was carried on following the semi-structured face to face interview and structured close-ended questionnaires' via Google Form. During this process several things related to the system requirements were explored as (a) how they monitor individual soldiers from mental health point of view; (b) how they are taking decision basing on those issues; (c) are they using any IT gadgets in fields to measure and check health issues of soldiers in active battle field; (d) how they feel the necessity of exploring such kind of technology to enhance the monitoring health of soldiers in fields; (e) do they feel that this tool will be useful for increasing accuracy of decisions in battle field; (f) whether soldiers will feel well about a tool being attached with their helmet; (g) how they will feel being monitored; (h) will that increase their confidence or make them scared of being creative. All the discussion were recorded instantly and documented later on. Furthermore, the transcript data along with the responses of Google form were being analyzed through content analysis and following significance of the stress management system named "Intelli-Helmet" were found which has been framed into a conceptual framework of the proposed system as shown in Fig. 1.

1. Such kind of system is really important and it will definitely boost confidence of soldiers during intensive military operations.
2. Generally no IT gadgets is being used to monitor the mental health status of the soldiers. However, automatic data capturing of brain waves and other physiological signals such as heart rate, Heart rate variability (HRV), blood pressure were appreciated.

3. Prior notice before crossing threshold was highly recommended for assisting the decision regarding continuation of deployment.
4. Auto processing of data to generate high value report were desired which can ease the decision making process.
5. Real time data visualization was expected for an overview of the stress factors by the commander to dominate on actual scenario.
6. Storage of mental health was to be maintained meticulously. Moreover, the concept of tracing stress pattern and combat efficiency of individual soldier were highly appreciated.

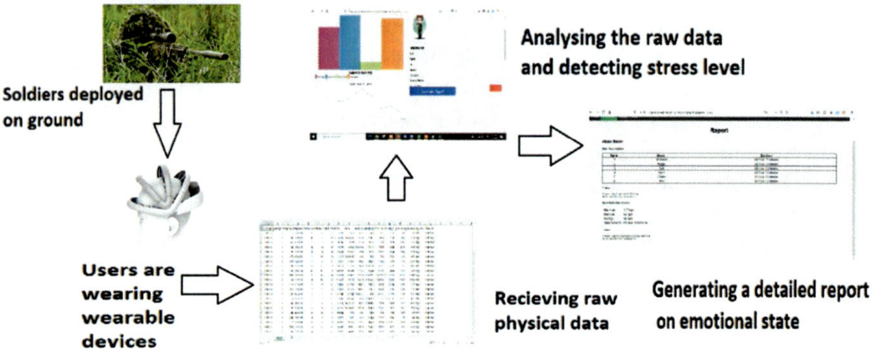

Fig. 1. Conceptual framework of the proposed system "Intelli-Helmet"

4 Design and Development

4.1 Data Acquisition

Based on the conceptual framework, system architecture of the proposed system named "Intelli-Helmet" was planned as shown in Fig. 2.

For collecting the readings of brain wave we have used Electroencephalography (EEG) which is a non-invasive method of obtaining brain waves. Neurosky Mindset device, which provides a single channel of EEG recording from a dry electrode placed at the frontal location (FP1) of brain, referenced to an electrode placed at the ear lobe was used to collect EEG. The device was placed inside the Military helmet as such so that it can collect the brain waves of a soldier in battlefield. It was configured at a sampling rate 512 Hz to record discrete EEG data. Mi band-4, which is an embedded system to monitor various physiological parameters was used to collect Heart rate (HR), Heart Rate Variability (HRV) etc. Data was collected using android applications named "EEG-Id" for Neuromind and "Mi-Fit" for Mi-band 4. Lastly, using the customized application of "Intelli-Helmet" CSV format of raw data were produced and fed as an input to the main system.

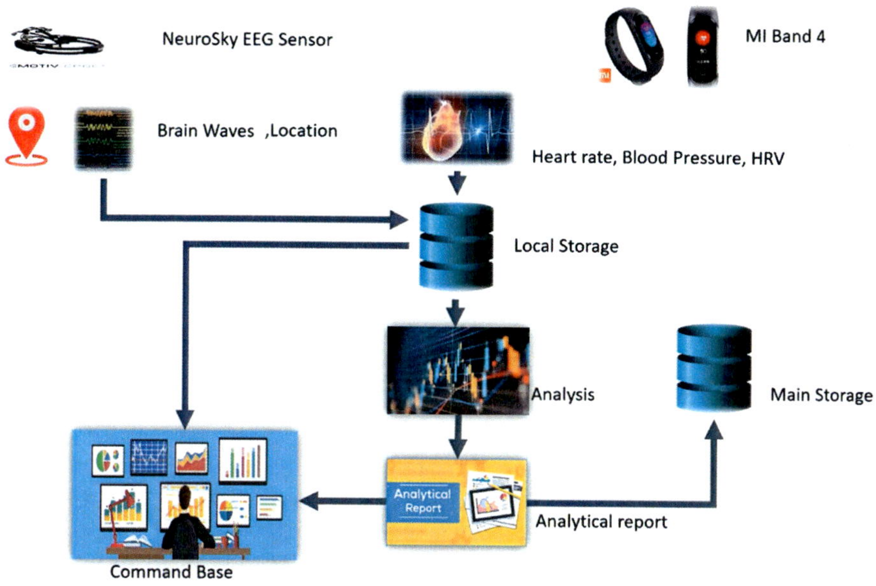

Fig. 2. System architecture of the proposed stress monitoring system

4.2 Model Construction

Data of brain waves and other physiological parameters were collected using the system. After noise cancellation and pre-processing total 6318 labeled data with a dimension of 486 rows and 13 columns were taken into consideration for analysis. For evaluating the machine learning algorithms data mining system WEKA was used and six state-of-the-art classification models (Bayes Net, Naive Bayes, J48, JRip, AdaBoost and Random Forest algorithm) were used for predicting stress status of the participants.

We used Accuracy (Kappa-statistics), precision, F-score, Sensitivity and specificity as evaluation parameters and test prediction models using Percentage-Split train/test (70/30) separation. Table 1 gives a comparison of performance parameters for the selected algorithms while predicting the stress status basing on the given data-set. When evaluating using Percentage-Split train/test (70/30) separation, we observed an overall better performance from Bayes Net and J48 decision tree (shown in Fig. 3). In case of accuracy (Kappa-Statistics) and precision Bayes Net and J48 show a better result whereas ADABoost and JRIP were less significant. Likewise, in case of F-score we observed a dominating performance from Bayes Net and J48 decision tree. However, while analyzing Sensitivity we observed a better performance from Bayes-Net, AdaBoost and JRIP than J48 decision tree whereas in case of Specificity we observed excellent performance from AdaBoost and JRIP with Bayes net in the second position. Hence, overall Bayes Net gives the best performance with an accuracy of 97.4% among all classifiers with J48 decision tree a close second with 96.1% accuracy.

Table 1. Evaluation parameters for stress detection using percentage split 70/30

	Accuracy (K-S)	Precision	F-score	Sensitivity	Specificity
Random Forest	0.9301	0.965	0.965	0.965	0.973
AdaBoost	0.9533	0.978	0.977	0.977	1
Naive Bayes	0.8759	0.941	0.939	0.939	0.974
Bayes Net	0.974	0.987	0.987	0.987	0.989
J48	0.9613	0.982	0.981	0.981	0.966
JRip	0.95	0.976	0.975	0.975	1

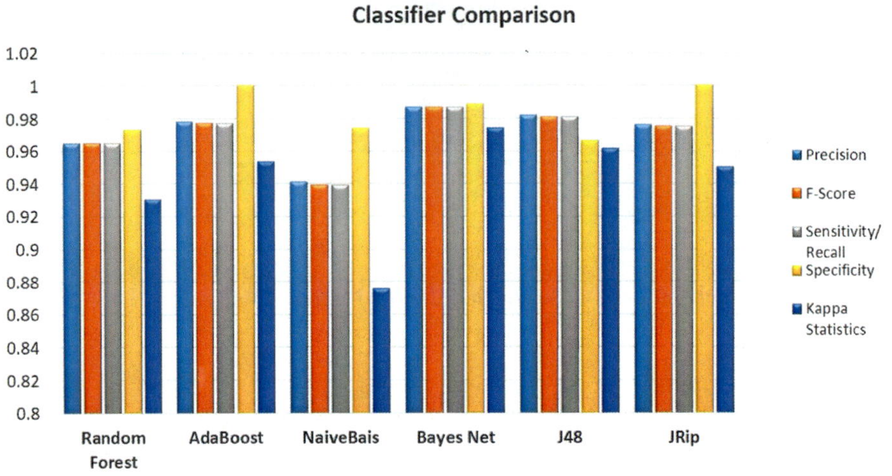

Fig. 3. Results of stress status classification using percentage-split (70/30) for different classifier models

4.3 Development of User Interface

The fetched dataset is then uploaded into a google sheet which can be used in any device using the web-based Graphical User Interface(GUI). The GUI is connected with a database in a local server which updates itself each time the system is used. When a new data is input from user after it is stored in google sheet, it goes through an anaconda program basing on finding out the stress level using a pre-constructed Bayes-net model and stored in the central database of the system. Then the data is fetched from the database and various parameters are shown through the system GUI. As shown in Fig. 4a Various battle-groups can enter into the system allowing a broad vision from top hierarchy command. Data visualization of various stress factors with respect to the threshold value along with personal information in individual profile is portrayed in real-time as shown in Fig. 4b. Moreover, their history is being recorded at the back-end to keep track of the stress pattern. A notification will be generated automatically with the detection of the subject as "STRESSED" and will suggest personnel

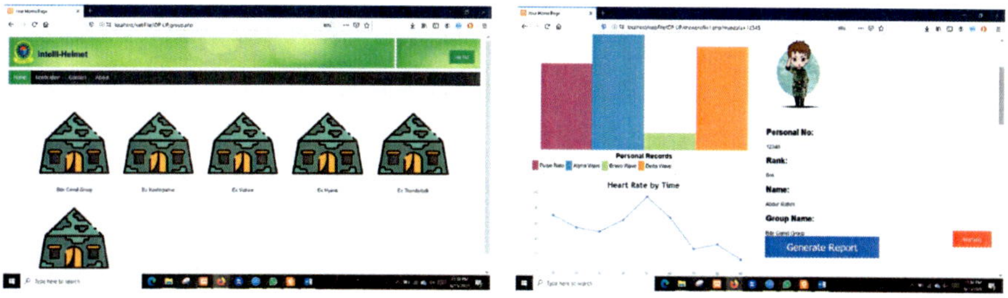

(a) List of Ongoing Operations

(b) Profile of Individual Soldier

Fig. 4. UI design of Inteli-Helmet

name who can replace the deployed soldier to achieve quick response in complex battle scenario. After generating the notification the report will be procured basing on the analysis of the stress factors. Main findings with summarised graphical representation along with recommendations will be generated to assist precise decision and planning.

5 Functionality and Usability Evaluation

The functional accuracy and usability performance of the projected stress monitoring system, "Intelli-Helmet" was evaluated through a demonstration exercise named as "Exercise Vulture" at Software Testing laboratory of authors' institute to validate the efficiency of the system using real-time data from user.

5.1 Participants' Profile

A total of 11 military personnel were invited as test-participants. Their age was between 23 and 35 years. Service ranks were varied from Lt Col (1), Major (2), Lieutenant (4) and Other Ranks (4). Service Experience of the participants was average 8.5 years. All of them had participated at least five military operation or had completed minimum 10 military exercises.

5.2 Study Procedure

At first, the participants were briefed to give an idea about the purpose of the study and their role during the study. Later, a short presentation on the Intelli-Helmet followed by a demonstration of using the system was provided. After that, each participants used the system for 5–10 min to get adapted with the system. Then the participants participated in the exercise followed by an interviewing session to gather information about their experience of the system. To conduct the evaluation study following steps were followed: a) One of the participant acted as an operation commander of Op and other 10 participants (with the Intelli-Helmet device) were asked to take one of the specified role of Operation commander at High Value Target(HVT) or Sniper section commander; b) Commander asked participants to take their specified role and observe their brain waves and physiological signals from control panel in a time span of 06 h; c) Participants were exposed in specified role based physical and mental exhaustion; d) Commander observed the stress factors of each participant and their stress status (notification while crossing the threshold) through the web-based application for the entire duration of the field study; e) Reports generated were displayed to the commander to assist the decision of deployment for an individual under-command (Fig. 5). On completion of the exercise, participants were asked to share their opinion about the usefulness and usability of the Intelli-Helmet system.

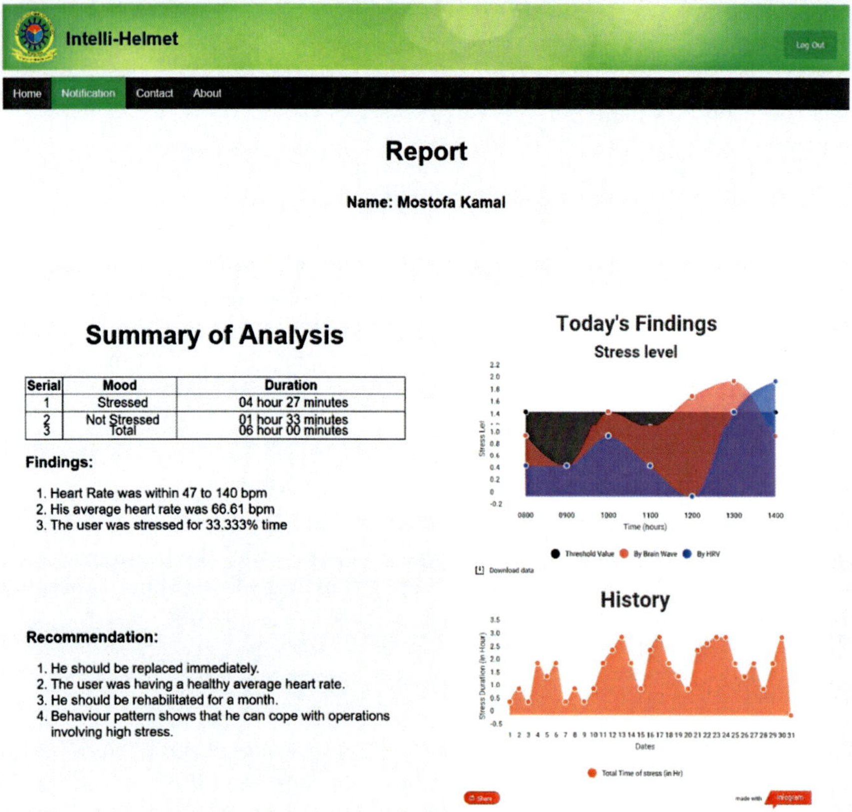

Fig. 5. Generated report from "Intelli-Helmet" for decision assistance

5.3 Results

Among 10 participants of specified role in under-command 70% (7 out of 10) were detected stressed and 30% (3 out of 10) were stable. The system accurately classified the stress status 100% participants (10 out of 10).Moreover, commander could assisted successfully to decide deployment of 90% participants (9 out of 10) in the specified roles. In a similar note, the web-based application showed alarm notification for 100% stressed participants. The interview results showed that all participants found the device interesting, useful, and easy to learn and use. They were also interested to recommend this device to their colleagues and to use the system in future. They highlighted & appreciated the issues of the cost effectiveness and the portability features. Some of the participants suggested to improve the power consumption of the EEG machine inside the helmet and also suggested to make it more comfortable.

6 Conclusion

The outcomes of this research can be discussed in several perspectives. First, the research developed an early prototype of a stress monitoring system which is suitable for military operations. The proposed system monitor the real time data and visualize the stress factors graphically along with the stress status outcome. The proposed system collect data related to the different stress factors during a military operation like Alpha and Beta brain wave, Heart rate, HRV and Blood Pressure, and then analyze the collected data through classifier. Secondly, based on the different classifying algorithm, the Bayes net algorithm showed the best performance with 97.4% in terms of accuracy while J48 decision tree is in a close second position with an accuracy of 96.1%. Finally, the evaluation study showed that the proposed system is useful, efficient and effective to perform its functionality during any military operation. Participants also expressed good level of satisfaction on the system basing on the efficiency and user experience. Furthermore, they have acknowledged that the report generated from the system through a collective analysis of stress factors will help higher command to take accurate decision in battlefield scenario.

The research has a few limitations as well. The evaluation study was conducted in an indoor environment and the number of participants were not adequate. The future work will focus to develop the concert system. Similarly, the updated system will be assessed by sufficient number of participants (or soldiers) in different terrain and real time Operational environment through field study to justify its effectiveness. Potential future research may be conducted focusing to assess the applicability and effectiveness of the proposed system during the military training sessions.

References

1. O'Neil, M., et al.: Factors associated with mild traumatic brain injury in veterans and military personnel: a systematic review. J. Int. Neuropsychol. Soc. **20**(3), pp. 249–261 (2014)

2. Bryan, C.J., Clemans, T.A.: Repetitive traumatic brain injury, psychological symptoms, and suicide risk in a clinical sample of deployed military personnel. JAMA Psychiatry **70**(7), 686–691 (2013)
3. Choi, K.H., Kim, J., Kwon, O.S., Kim, M.J., Ryu, Y.H., Park, J.E.: Is heart rate variability (HRV) an adequate tool for evaluating human emotions? - A focus on the use of the international affective picture system (IAPS). Psychiatry Res. **251**, 192–196 (2017)
4. Razzak, M., Islam, M.N.: Exploring and evaluating the usability factors for military application: a road map for HCI in military applications. Hum. Factors Mech. Eng. Defense Saf. **4**, 4 (2020). https://doi.org/10.1007/s41314-019-0032-6
5. Diya, S.Z., Prorna, R.A., Rahman, I.I., Islam A.B., Islam, M.N.: Applying brain-computer interface technology for evaluation of user experience in playing games. In: 2019 International Conference on Electrical, Computer and Communication Engineering (ECCE), Cox's Bazar, Bangladesh, pp. 1–6 (2019)
6. Islam, M.N., Islam, M., Islam, S., Bhuyan, S.,Hasib, F.: LocSoldiers: towards developing an emergency troops locating system in military operations. In: 2018 4th International Conference on Electrical Engineering and Information & Communication Technology (iCEEiCT), Dhaka, Bangladesh, pp. 264–267 (2018)
7. Vaid, S., Kaur, C., Singh, P.: EEG analysis for emotion recognition using multi-wavelet transform. Int. J. Pharm. Technol. IJPT. **9**(1), 29222–29229 (2017)
8. Abdulkader, S.N., Atia, A., Mostafa, M.S.M.: Brain computer interfacing: applications and challenges. Egypt. Inf. J. **16**(2), 213–230 (2015)
9. Gonfalonieri, A.: A beginner's guide to brain-computer interface and convolutional neural networks (2018). https://towardsdatascience.com/a-beginners-guide-to-brain-computer-interface-and-convolutional-neural-networks
10. Best, J.: What is a brain-computer interface? Everything you need to know about BCIs, neural interfaces and the future of mind-reading computers. Article, Building the Bionic Brain (2019)
11. Murugappan, M., Ramachandran, N., Sazali, Y.: Classification of human emotion from EEG using discrete wavelet transform. J. Biomed. Sci. Eng. **3**(4), 390–396 (2010)
12. Sriramprakash, S., Vadana, P.D., Murthy, O.V.R.: Stress detection in working people. Procedia Comput. Sci. **115**, 359–366 (2017)
13. Bos, D.P.: EEG-based emotion recognition. The influence of visual and auditory stimuli (2006)

Assessing the Performance of Online Food Delivery (OFD) in India

Amogh Bhaskara[1], Siddharth Menon[1], U. Dinesh Acharya[1], and H. C. Shiva Prasad[2](✉)

[1] Manipal Institute of Technology, Manipal Academy of Higher Education, Manipal 576104, India

[2] Department of Mechanical Engineering, Faculty of Engineering, School of Automobile, Mechanical and Mechatronics, MUJ, Jaipur 303007, India

shiva.prasad@jaipur.manipal.edu

Abstract. The purpose of this paper is to analyze the application quality of the online food delivery companies by using a qualitative and exploratory approach through the collection of primary data through a consumer survey and secondary data through a sample of 25 companies operating in the online food delivery sector in India. The analysis has been conducted through the parameters involving the aspects of Content, Usability, Functionality, Trust, Satisfaction and Brand Loyalty. The Data gathered has been analyzed using the SPSS followed by the Structural equation modelling (SEM). The conceptual model for the project has been achieved using the SmartPLS (v.2.3.8). The result shows better usability to the customer has a link to enhance application quality for business cycle and loyal customer trust the OFD system to allow efficiency building and look for profiteering space.

Keywords: Online food delivery · Trust · Brand loyalty · Application quality · Usability

1 Introduction

With the rise in the e-commerce industry in India, online food ordering, and delivery services have had tremendous growth. It has changed the dynamics of the food industry from brick-and-mortar restaurants to delivery of food to the customers' doorsteps. The shift of attention towards online food delivery (OFD) services is associated with the rise in the income, consumption levels, favorable lifestyle changes, cheap, widespread access to the internet, the convenience of online ordering, and aggressive marketing strategies adopted by online food startups [1]. With the increasing number of working women in India, they tend to focus on productivity and spend their time either commuting and working, as they have very little time for cooking food at home. The customers enjoy the array of benefits, but the restaurants have scope for a lot of growth potential, especially in terms of orders and the amount of revenue generated that has helped many businesses to expand their geographical locations [2]. The commercial food ventures were able to

L. Garg et al. (Eds.): ISMS 2020, LNNS 303, pp. 33–45, 2022.
https://doi.org/10.1007/978-3-030-86223-7_4

simplify tasks and conserve on investments in non-core affairs. In the meantime, for these brands, the increased use of the services by an ever-growing number of consumers has caused higher influx of accesses, an element that serves as an indexing factor to attract new restaurants to its website/application and to attract investors to aid in the expansion of the business to other regions [3]. The rate of increase of these OFD services into smaller towns and areas has surged significantly in 2019 [4]. The business model used by OFD services that were once seen as a 'metro phenomenon,' limited to the metro cities of India, is now rapidly accounting several record numbers of hits in tier two cities.

The goal of these platforms is to act as an intermediary service linking final consumers and commercial establishments. Hence, transparency in communication and adaptation to the selected audience's needs are indispensable for sustaining the business [4]. The statistical reports by RedSeer research and consulting group, the food-tech market of India indicates that the increasing order frequency by the consumers is the critical leveraging factor in the food-tech sector's growth in the past few years [4]. The current top market players, Swiggy, and Zomato are recording approximately 30–35 million orders each month. Hence, the keynote is that the customers have shown increasing interest and support towards the online food delivery applications [5]. For the food-tech market of India to maintain a steady presence, it needs to focus on improving both customer brand loyalty as well as customer acquisition rate by ensuring customer trust. Both the parameters significantly influence the OFD supply chain and the value chain.

Being a relatively new service industry in India, the growth of OFD companies is complemented by a stable sector intermixing and by the conception of large industry players and aided by foreign investments. With the exponential growth of this relatively new market segment in India, it has become pivotal in determining the characteristics of the market competitors and in analyzing their performance. In principle, this sector has brought benefits along with formidable challenges. The arrival of specialized companies in providing OFD services has enabled the users to pick the product/service (restaurants, food choices on the menu, and payment methods) pronto and effortlessly compare between an array of listed options. Also, the inception of innovative business models has enabled these companies to infiltrate into the most remote markets within the country. Tracking these novelties and their impact on the ever-changing consumer market is essential. Thus, to address these problems, few research objectives are formulated in the following section.

The objectives of this paper are to characterize the performance of OFD companies present in India and to analyze the content of the websites/applications of these companies with a view to its utilization as a site for conducting business transactions [5]. Finally, to analyze the OFD platforms from the selected sample list, based on three critical dimensions are content, functionality, and usability.

2 Literature Review

In the present scenario, the non-core market constituting more than 100 cities experiences 30% of the order traffic as compared to the top seven core cities that amass the major split of the order traffic [4]. It is vouched that since December 2018; the OFD services are striving to expand their geographic presence in multi-cities, making their presence known in 100 cities and more. Though there is a significant metro cities hub that retains most of the order traffic are from New Delhi, Chennai, Mumbai, Kolkata, and Bangalore. The shift towards the non-core market is taking place with an awareness of e-commerce. There are pockets of experimentation with multiple online-offline models to tap into the next growth opportunities more effectively. Many such marketing campaigns are positively impacting the profit-margins and ensuring sustained progress [6]. With so many innovations being brought about on these platforms, many cross-platform business models are emerging to change the market dynamics. To name a few noteworthy innovative programmes, infrastructure set-up availed by restaurants through Swiggy Access, restaurant raw-material supply chain offered by Zomato's HyperPure, collaborations with NGOs to fight food wastage by Zomato, privileged customer programs like Swiggy Super and Zomato Gold and others. A groundbreaking and innovative business model, HyperPure by Zomato aids to deliver transparency and accountability over the food supply value chain. HyperPure has tie-ups with several vendors and farmers who provide high-quality raw materials to their warehouses, which in-turn supplies to about 2500 restaurants within Bangalore, who are listed on their website/application. This provision of clean and high-quality ingredients to restaurants reinforces the efficiency of the food supply chain by involving the minimal number of certified parties in the process. The Swiggy Access provision offered by Swiggy to their restaurant partners is another noteworthy business model that is observed to have gained much traction. Swiggy makes use of the data extracted from their extensive databases and generates reports, focusing more on the gaps in the supply-chain on a highly localized level. Once these gaps are identified, they move on to invite their partner restaurants to reckon the demand of the local residents. The model can be interpreted as a highly mutually benefitting one that enhances the operations of the existing restaurant partners by enabling them to initiate expansion and simultaneously satisfies the accessible cuisine needs of the local population. The aggregator fetches themselves a gradually growing revenue model by levying 3–8% per order value. Reports suggest that Swiggy is already recording increased confidence in the model from their restaurant partners, and almost 70% of the partners are discussing the setting up at a second location [4]. Keeping the end-users always engaged with valuable offers and schemes is key to building strong customer bases. The customer trust is being captured by rolling out value-added services like Zomato Gold and Swiggy Super that grants the customers freebies and incentives with the additive use of their services. The establishment of such varied business models balances the gap between the heavy cash burnouts and the gradual increase in revenues of the OFD competitors. The Indian food-tech sector is yet to enter the second sage also known as turbulent stage and is forecasted to experience in smaller towns than the other areas [4, 5]. It is undoubtedly going to be extremely exciting to see how the Indian food-tech market pans out and takes shape, as it is already one of the fastest-growing markets among other global industries.

3 Methodology

The primarily conceptual model chosen was a flowchart that bridged the multiple secondary factors regarding application quality and service quality with customer loyalty. For this project, the exploratory and qualitative research approaches were followed through the collection and analysis of data on OFD companies registered in India. Reviews of their websites/applications were also conducted by investigating parameters proposed [6], involving aspects such as content, usability, and functionality. The parameters used to represent the criteria, as mentioned in [6] as adopted by [7], later refined by [8] is taken for the conduct of the analysis of OFD services in Brazil [8]. The analysis procedure is replicated in this project to form a better understanding of the Indian food-tech market.

For the definition, web searches were conducted using keywords such as online food delivery, fastest food delivery, local food delivery, order food online, and order from the mobile phone. A list of companies is devised, using reliable information observed in newspaper articles, magazines, blog-spots, and websites of companies offering OFD services in India [9]. The sample from the list of 25 companies, in correspondence with the survey summer period of three months. The analysis of the websites/applications of the competitors was surveyed by focusing on the observation of the subsistence of the parameters that make up the dimensions content, usability, and functionality. Thus, when the evaluation of criteria for a projected parameter was observed, yes was filled, and when its absence was noted. The sample data gathered in our survey is used to analyze the online food delivery applications in India. Also, a set of statistical reports on the performance of these competitors are extracted. The validation and analysis of the result data from the survey were done utilizing SPSS and SmartPLS software.

The operational definition of the constructs and the variables renders a clear understanding and aids in avoiding the susceptibility to interpersonal influences. The criteria and parameters, as explained [8], are listed in the following sections.

3.1 Content

The construct content accommodates three unique parameters that define the quality, compatibility, reliability, and reach of the content of the website/application. The first parameter scope, coverage, and purpose checks for links to other information sources and the social media presence of these websites/applications. The second parameter objectivity studies the relevance of content is displayed clearly and consistently on the website/application. The third parameter authority goes over the legitimacy of the website/application offering services [10].

3.2 Usability

Much of the web content is usable and used frequently. In respect to the dimension usability, the parameter operability is probed by looking at the layout adjustment capabilities of the website/application to fit into varying resolutions on multiple devices. Device-optimized websites/applications are preferred by the end-users as it is easier to navigate. The second parameter learnability works towards identifying a broad set of features of the website/application, including the availability and findability of navigation/search tools, added information indicators, and human interface for assistance. The third parameter intelligibility verifies that each link carries out a unique function and that it is adequately named.

3.3 Functionality

Subsequently, for the construct functionality, the security parameter was audited based on the information available in the form of the privacy policies in portals, such as cookie encryption capabilities and the use of secure/well-established protocols for their websites. The parameters interoperability and accuracy were analyzed either by placing simulated food delivery orders on-line or by extracting genuine user experience reviews to identify information on payment methods offered, delivery fee values, time of execution of the request, and evaluation of the quality of service, provided by the end-users. Finally, through the adequacy parameters, the extent of reciprocating dialogue taking place between the companies and customers was analyzed through comments, general query sections, compliments/suggestions received by companies through social networks such as Facebook, Twitter, Instagram, and blogs.

The percentage distribution of the dimensions and parameters was obtained using descriptive statistics among the 25 websites/applications analyzed. Assigning precise weights to the parameters is an arduous task as the above-described analysis of the websites/applications is not adequately refined, and the objectives may deviate. One of the explanations is that the results of the averages for the three dimensions might result in the distortions between some parameters. When there is a distortion in the results, it is indicated in the descriptive analysis. A survey conducted by a questionnaire that skimmed the customer's point of view on the criteria and parameters. Questions were devised such that the answer to each question validated parameters. The data gathered in our pilot survey was used to gain better insight into the chosen criteria. Variance-based structural equation modeling (SEM) that uses the partial least squares path modeling method was used to get the results. Smart-PLS is a freeware/proprietary software that was used to conduct the variance-based SEM on the gathered data. Validation of the results procured was done by using SPSS software. The result and discussion of the analysis are covered in the next section. For a better understanding of the conceptual framework, a model is depicted in the next section to specify the specific direction of the relationship between the constructs and the parameters.

The secondary data for the pilot study was gathered through a survey. It provided insights into the customer perception of the OFD brands in terms of the criteria and parameters chosen for the study. A strong correlation was observed between the primary and secondary data collected. The total sample size of the data gathered is 353 respondents, out of which many of them belong to the Gen Z, i.e., the people who belong to the age group of 20–25 years. The Gen Z population are the power users of the OFD services; therefore, the demographic distribution of the sample population is justified. Table 1 gives the demographics of the sample population.

According to the data obtained from the survey, Swiggy and Zomato are the most popular OFD brands among the sample population, followed by Dominos, UberEats, and McDonald's, respectively. Because most of the sample population resides in the Udupi/Mangalore region of India, the regional OFD services, Foodzoned.com and Foodzozo.com, are also among the popular services listed [11–14]. Such a discrepancy in the demographics of the sample population will not affect the result analysis, when viewing the issue from a holistic approach, citing the problem applies to the India companies.

Table 1. Demographic profile of respondent.

Field	Category	Respondent
Gender	Male	253 (71.70%)
	Female	100 (28.30%)
Age	15–20	43 (12.18%)
	21–25	284 (80.45%)
	26–30	14 (3.97%)
	31–35	3 (0.85%)
	36 and above	9 (2.55%)
Profession	Full-time employed	80 (22.70%)
	Part-time employed	26 (7.40%)
	Unemployed	10 (2.80%)
	Full-time student	219 (62.00%)
	Part-time student	10 (2.80%)
	Others	8 (2.30%)

4 Results

The analysis of the study is obtained through both descriptive and inferential statistics. The factor analysis is done to examine the pattern of correlation between the observed measures. The data gathered is analyzed using the SPSS followed by the Structural Equation Modelling (SEM). The conceptual model developed with SmartPLS (v.2.3.8). The research model is analyzed and interpreted in two stages. The first stage consists of the assessment and refinement of the adequacy of the measurement model, and the second stage consists of the assessment and evaluation of the structural model.

To understand the various characteristics of the data and helps in understanding the procedures in hypothesis testing. The study looks at the reliability of the scale used. It helps us determine whether the scale used is reliable and can support in explaining the data. The study employs Cronbach's alpha coefficient to track the internal consistency of the scale. Author like Pallant [15] advocates above 0.7 construct factor value is a valid measurement. The Cronbach's alpha value calculated for the conducted survey sample data and the function of the number of test items and the correlation among the items are also listed in the following section [19] and the formula used for Cronbach's alpha.

$$\alpha = \frac{\mathrm{Nx}\underline{\mathrm{c}}}{\underline{\mathrm{v}} + (\mathrm{N}-1)\mathrm{xc}} \tag{1}$$

Where, N is the number of items,

c is the average inter-item covariance among the items and,

v is the average variance.

Cronbach's Alpha test is used to understand and calculate the reliability of the survey questionnaire. It was obtained as 0.899 that ensures that the items measured are consistent [19]. This is followed by factor analysis that helps to understand independent variable has the most substantial influence over the others. The Cronbach's Alpha will generally increase the correlations between the items [20]. The maximum value is one, and minimum value is zero. The reading is expected to be in the range between 0.70–1.00.

The PLS model in Fig. 1 is interpreted in two stages. In the first stage, the assessment and refinement of the measurement model are carried out, followed by the formation of the structural model. The coefficient of determination is R^2 0.675 for the application quality endogenous latent variable. The coefficient of determination R^2 is 0.555 for the endogenous latent variable, trust. The coefficient of determination R^2 is 0.521 for the brand loyalty endogenous latent variable. The optimum value range for the coefficient of determination, any value above a minimum of 0.25 [17].

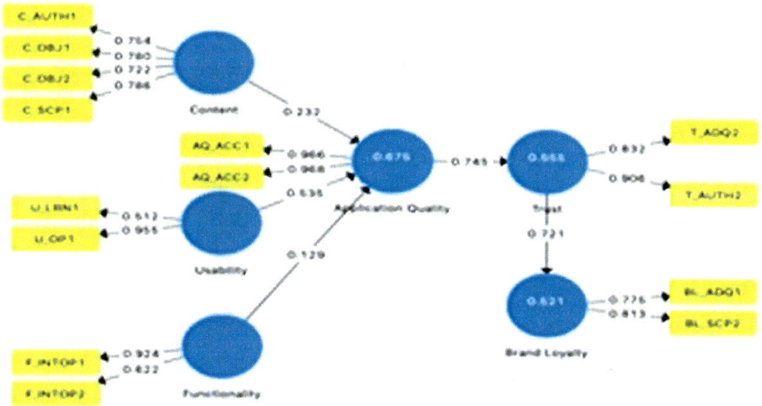

Fig. 1. The PLS algorithm Structure Equation results

This leads to the conclusion that the latent variables (content, usability, and functionality) have the following path coefficients (see Fig. 2).

The chart of path coefficients of the latent variables in indicates the Application Quality relates to Trust has the highest path coefficient, which suggests that the application quality has the highest affinity towards the trust of the user.

The Usability relates to Application Quality has a path coefficient of 0.535 that is much higher than Content relating to Application Quality and Functionality relates to Application Quality. This indicates that the usability of the website/application contributes more to the application quality than what, content or functionality adds to the application quality of the website/application. The lowest path coefficient noted is Functionality relates to Application Quality, a value of 0.129 which indicates that the users have given the functionality factor of the OFD website/application the least consideration/importance.

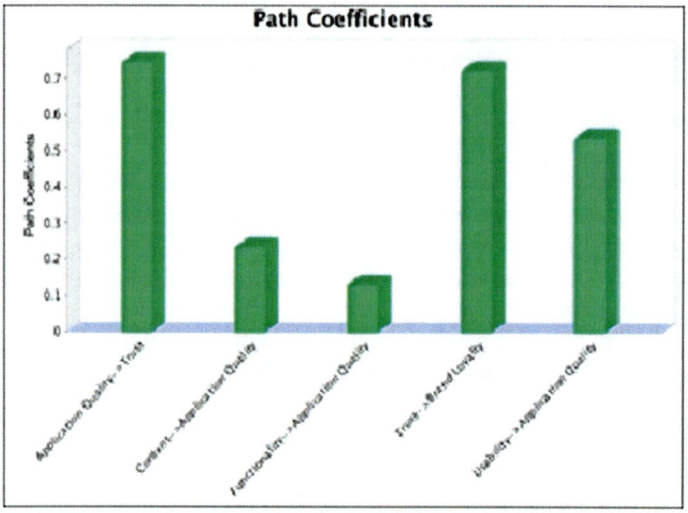

Fig. 2. Path coefficients of the Latent Variables

The test for validity and reliability of the constructs, as mentioned in the procedure for the development of the scale has have correct measure [21]. The indicator reliability test, and then the convergent validity tests and discriminant validity test is conducted if the given scale is evaluated (see Table 2).

Table 2. Indicator reliability

Item	Indicator	Loading	Reliability
1.	C_OBJ1	0.780	0.6084
2.	C_AUTH1	0.754	0.568
3.	C_SCP1	0.786	0.617
4.	BL_ADQ1	0.793	0.628
5.	T_ADQ2	0.832	0.692
6.	F_INTOP1	0.924	0.853
7.	U_OP1	0.955	0.912
8.	AQ_ACC1	0.966	0.933
9.	AQ_ACC2	0.968	0.937
10.	F_INTOP2	0.924	0.853
11.	~~T_AS1~~	~~0.224~~	~~0.050~~
12.	BL_SCP2	0.796	0.633
13.	T_AUTH2	0.983	0.966
14.	~~T_ADQ4~~	~~0.343~~	~~0.117~~
15.	C_OBJ2	0.722	0.521

Like other marketing research procedures, it is vital to declare the reliability of the latent variables to fulfill the examination of the structural model. The indicator reliability value of the latent variable is calculated by taking the squared value of all the indicator loading values. For research purposes, it is requisite to have an indicator reliability value higher than 0.70. In the case of exploratory research, an indicator reliability value higher than 0.40 is preferred [22].

The lowest path coefficient noted is Functionality relates to Application Quality, a value of 0.129 which indicates that the users have given the functionality factor of the OFD website/application the least consideration/importance.

4.1 Convergent Validity

Convergent validity specifies that items that are indicators of a construct should share a high proportion of variance [21]. The convergent validity of the scale items was assessed using the given criteria and the factor loadings should be higher than 0.50 [21]. Secondly, the composite reliability for each construct should exceed 0.70. Lastly, the average variance extracted (AVE) for each construct should be above the cut-off 0.50 proposed by [21, 22].

4.2 Discriminant Validity

The Criterion Analysis (CA) that is opted for checking discriminant validity tells that the square root of AVE in each latent variable is used to establish discriminant validity [20]. The value must be larger than the other correlation values among the latent variables and the square root of AVE was manually calculated and diagonally bold represented (Table 3).

Table 3. Discriminant validity

	AQ	BL	CONT	FUNCT	TRUST	USAB
AQ[a]	**0.967**					
B L	0.594	**0.794**				
CONT	0.688	0.536	**0.761**			
FUNCT	0.707	0.562	0.721	**0.788**		
TRUST	0.745	0.721	0.711	0.718	**0.870**	
USAB	0.819	0.617	0.732	0.770	0.778	**0.742**

[a]AQ = Application Quality; BL = Brand Loyalty; CONT = Content. FUNCT = Functionality; USAB = Usability.

The correlations between the latent variables were transferred from the Latent Variable Correlation. For example, the latent variable, functionality has the AVE found to be 0.621 (from Table 3). Hence, its square root becomes 0.788. This number is larger than the correlation values in the row of functionality (0.707, 0.562, 0.721 and 0.650) and more significant than those in the column of functionality (0.718 and 0.770). Similar observations are noted for the latent variables, application quality, brand loyalty, content, functionality, trust, and usability. The result obtained indicates that the discriminant validity is well-established as all the values calculated along the principal diagonal of the LCA matrix confirms to the required criteria.

4.3 Discussion

The path coefficient for the model is carried out. The review aids in exploring the outer model by checking the T-statistic in the Outer Loadings (Means, STDEV, T-Values). The t-statistics are more significant than 1.96 that the outer model loadings are highly significant (Table 4).

Table 4. Path coefficient and significance (t-stat)

Application quality to trust	23.918^{***}
Trust to brand loyalty	26.211^{***}
Content to application quality	4.832^{***}
Functionality to application quality	2.378^{**}
Usability to application quality	10.922^{***}

$^{***}p \leq 0.001$, $^{**}p \leq 0.01$, $^{*}p \leq 0.5$.

A hypothesis testing is conducted to foresee and evaluate the result criteria by forming a hypothesis and using the t-statistics value for each of the path loadings. The higher the t-statistics value indicates that the higher significance support to the path, median, standard deviation, and mean. The cut-off criteria used was value greater or equal to 1.645 for an alpha level of 0.05 [17].

The research hypothesis H1 stated that there is a positive relation between application quality and trust. The hypothesized path for H1 was positive ($t > 1.645$ and $p < 0.01$) and is supported. Thus, it can be said can say that the null hypothesis is accepted, and an alternate hypothesis is rejected. The second research hypothesis H2 states that there would be a positive relationship between content and application quality. The hypothesized path for H2 was positive ($t > 1.645$ and $p < 0.01$) and thus, the null hypothesis is accepted, and an alternate hypothesis is rejected.

Hypothesis H3 states that there is a positive relation between functionality and application quality. The hypothesized path for H3 was positive ($t > 1.645$ and $p < 0.01$) and is supported. Thus, the null hypothesis is accepted, and the alternate hypothesis is rejected. Hypothesis H4 states that there is a positive relationship between trust and brand loyalty. The hypothesized path for H4 was positive ($t > 1.645$ and $p < 0.01$) and is supported. Thus, the null hypothesis is accepted, and the alternate hypothesis is rejected. Hypothesis H5 states that there is a positive relation between usability and application quality. The hypothesized path for H5 was positive ($t > 1.645$ and $p < 0.01$) and is supported. Thus, the null hypothesis is accepted, and an alternate hypothesis is rejected. The model helps in understanding the intricacies of the OFD structure, study on how it benefits the application quality when it is compared with the path coefficients and how it affects each other differently. The model was assessed using three criteria: 1) path coefficients (β), 2) path significant (p-value) and 3) variance explained. The model was designed in PLS as per the guidelines are given in the SmartPLS Guide [17].

The criterion put forth by Rossiter [23] states that for the structural model all paths should result in a t-statistic value greater than two and the latent variable' R-Square values are higher than 50%. The result for the full model indicates that 62.7% of the variance in application quality and 52.1% of the variance in brand loyalty. 55.5% of the variance in trust. Accordingly, from the structural model, all the predicted paths are hypothesized are validated. The model indicates that trust is having a positive impact on brand loyalty.

5 Conclusions

The dimensions presented, the usability dimension has the highest occurrence, followed by content and functionality. In the analysis of the path coefficient model, the application quality dimension leading to trust has the highest path coefficient and indicates that with an application which consists of appropriate content, functionality and usability ensures that the consumers display a higher trust towards the brand. Through several tests conducted on the data gathered, it was possible to verify the reliability and credibility of the data collected. The OFD industry is still in its infancy and yet to enter the next stage. With time, the consequences and/or fruits of the services rendered may be assessed. In the current scenario of the Indian OFD services' market, it is very evident from the analysis conducted that the data suggests that there is a shift in the lifestyle of the population. Prolonged brand loyalty may lead to the consumers' fixating on a brand, and this leads to changes in the way the consumer perceives his/her intent to purchase products or services. Erecting a seamless experience for the consumers with stringent marketing efforts and creating innovations in the lifestyles of the general population should be the prime objectives of the OFD companies to be able to sustain for an elongated period. In the proceeding text, the future scope of the project is discussed.

The insights from this study may help the OFD companies in India to take note of the impacts their strategies are making, especially when it comes to interactions with the customers to enhance brand loyalty [24]. The brands will have to scrutinize the application quality of their existing services primarily and give high regard to the use of social media to boost their returns [25, 26]. Few limitations in this study since it is an exploratory form of research and the limited availability similar works from the literature. The data reliability has given confidence in transforming the little data to a meaningful analysis and interpretation. Future works gives an opportunity for probing on Online food sector. The dynamic nature of this booming market segment may intrigue the upcoming generations and encourage them to devise more innovative marketing campaign, and to produce changing ideas in the field of interest.

References

1. Gupta, A.: India hyperlocal market outlook to 2020 - Driven by Surge in Number of Startups and Series of Funding, kenresearch.com, para, 3 March 14 (2016). https://www.kenresearch.com/technology-and-telecom/it-and-ites/india-hyperlocal-market-research-report/7209-105.html. Accessed 28 Jan 2019
2. India Brand Equity Foundation, Growth of Ecommerce Industry in India-Infographic, October 2018. https://www.ibef.org/industry/ecommerce/infographic. Accessed 28 Jan 2019
3. AIMS Institute Pvt. Ltd., Online Food Service in India: An Analysis. AIMS, July 2017. https://theaims.ac.in/resources/online-food-service-in-india-an-analysis.html. Accessed 28 Jan 2019
4. RedSeer Consulting Agency, Food-Tech Market Updates, March 2019. https://redseer.com/newsletters/food-tech-market-updates-june18/. Accessed 19 Mar 2019
5. Meenakshi, N., Sinha, A.: Food delivery apps in India: wherein lies the success strategy? Strateg. Dir. **35**(7), 12–15 (2019). https://doi.org/10.1108/SD-10-2018-0197
6. Anumolu, K., Shiva Prasad, H.C., Gopalkrishna, B., Adarsh, P.K.: A digital marketing: a strategic outreaching process. Intl. J. Manage. IT Eng. **5**(7), 254–260 (2015)
7. Vilella, R.M.: Conteúdo, usabilidade e funcionalidade: três dimensões para an avaliação de portais estaduais de governo eletrônico na web, [Dissertação de Mestrado] Escola de Ciência da Informação, Universidade Federal de Minas Gerais, Belo Horizonte, 263f (2003). http://bogliolo.eci.ufmg.br/downloads/VILELLA%20Conteudo%20Usabilidade%20e%20Funcionalidade.pdf. Accessed 14 Feb 2019
8. Daim, T.U., Basoglu, A.N., Gunay, D., Yildiz, C., Gomez, F.:, Exploring technology acceptance for online food services. Int. J. Buss. Info. Syst. **12**(4), 383–403 (2013). https://www.inderscienceonline.com/doi/pdf/10.1504/IJBIS.2013.053214
9. Pigatto, G., Guilherme, J., Negreti, A., Machado, M.L.: Have you chosen your request? Analysis of online food delivery companies in Brazil. British Food J. **119**(3), 639–657 (2017)
10. Thongpapanl, N., Ashraf, A.R.: Enhancing online performance through website content and personalization. J. Comp. Info. Syst. **52**(1), 3–13 (2011)
11. Statista, Platform-to-Consumer Delivery Statistics, March 2019, Online statistics portal. https://www.statista.com/outlook/376/119/platform-to-consumer-delivery/india. Accessed 25 Mar 2019
12. Das, J.: Consumer perception towards online food ordering and delivery services: an empirical study. J. Mange. **5**(5), 155–163 (2018). http://www.iaeme.com/JOM/issues.asp?JType=JOM&VType=5&IType=5. Accessed 31 Jan 2019
13. Building Brands Online: Interactive Branding: Best Practices in a Direct Response-Driven Media, AdAge Insights (2010)
14. Kedah, Z., Ismail, Y., Ahasanul, K.M., Ahmed, A., Ahmed, S.: Key success factors of online food ordering services: an empirical study. Malays. Manage. Rev. **50**(2), 19–25 (2015). https://www.researchgate.net/publication/291074636_Key_Success_Factors_of_Online_Food_Ordering_Services_An_Empirical_Study. Accessed 31 Jan 2019
15. Pallant, J.: SPSS Survival Manual - A Step by Step Guide to Data Analysis Using SPSS for Windows, 3rd edn. Open University Press, Maidenhead (2007)
16. Srinivasam, S.: Swiggy pilots B2B offering under Swiggy Café at corporate cafeterias, ET Bureau, 3 October (2018). https://www.economictimes.indiatimes.com/articleshow/66048384.cms?utm_source=contentofinterest&utm_medium=text&utm_campaign=cppst. Accessed 21 Apr 2019
17. Wong, K.K.: Partial least squares structural equation modeling (PLS-SEM) techniques using SmartPLS. Market. Bull. **24**, 1-32 (2013). Technical Note 1. https://www.academia.edu/9210442/Partial_Least_Squares_Structural_Equation_Modeling_PLS-SEM_Techniques_Using_SmartPLS

18. Hair Jr., J.F., Hult, G.T.M., Ringle, C., Sarstedt, M.: A Primer on Partial Least Squares Structural Equation Modeling (PLS-SEM). Sage Publications, USA (2016)
19. Bruin, J.: newtest: command to compute new test. UCLA: Statistical Consulting Group (2006). https://stats.idre.ucla.edu/spss
20. Zaiontz, C.: Real Statistics using excel. http://www.real-statistics.com/reliability/cronbachs-alpha/
21. Fornell, C., Larcker, D.F.: Structural equation models with unobservable variables and measurement error: algebra and statistics. J. Market. Res. **18**, 382–388 (1981)
22. Tabachnick, B.G., Fidell, L.S.: Using Multivariate Statistics. HarperCollins, New York (1996)
23. Rossiter, J.R.: The C-OAR-SE procedure for scale development in marketing. Int. J. Res. Mark. **19**(4), 305–335 (2002)
24. Vos, K.: Ecommerce delivery trends: what contributes to a positive experience? (2016). https://www.mycustomer.com/selling/ecommerce/ecommerce-delivery-trends-what-contributes-to-a-positiveexperience
25. Alves, H., Fernandes, C.I., Raposo, M.: Social media marketing: a literature review and implications: implications of social media marketing. Psychol. Mark. **33**(1), 1029–1038 (2016). https://doi.org/10.1002/mar.20936
26. Krishnan, A.: Accessing the construct and content validity of uncertainty business using Sem approach - an exploratory study of manufacturing firm (2011). https://www.globaljournals.org/GJMBRVolume11/1-Accessing-the-Construct-and-Content-Validity-of-Uncertainty-Business.pdf

Incorporating Security Features in System Design Documents Utilized for Cloud-Based Databases

Rebecca Zahra and Joseph G. Vella(✉)

Faculty of Information and Communication Technology, University of Malta, Msida, Malta
{rebecca.zahra.10,joseph.g.vella}@um.edu.mt

Abstract. Cloud-based systems are being increasingly deployed due to their numerous benefits. Yet, there is uneasiness amongst organizations opting for such systems mainly due to security concerns. Security and protection of cloud-database systems from unauthorized access present countless challenges but are indispensable to address. Although security features should be included in the initial stages of system design, sometimes they are overlooked and left to the later stages in the development lifecycle. The framework proposed in this paper tackles this lacuna by including security directives at the initial design stage. It allows database designers to adorn the application's conceptual models namely entity structures, entity life history and data flow diagrams with security features. Discretionary and role-based access control mechanisms are utilized as the main form of security since they can counteract a high portion of security threats. The proposed framework consists of the creation of a unique security profile for each tenant and his users and an analysis algorithm which assists in the detection of possible security pitfalls. Based on the system's design data provided and security features encoded, this framework is then responsible for testing the overall design; for example, to ensure reachability and isolation of all database objects, functions and roles. Once the design, now supplemented with security features, is evaluated and deemed to be acceptable then SQL language constructs corresponding to the secure database design are generated. The framework is also useable when the cloud database goes live as any of the underlying security specifications can change during run-time, thus ensuring that security is always accounted for and manageable by developers and later on by the tenants themselves.

Keywords: Cloud systems · Multi-tenancy · Design diagrams · Data security

1 Introduction

The ever-growing dependency on computing pushes the constant introduction of novel technologies such as cloud computing. One significant change brought about by cloud computing is that related to database systems since companies are now utilizing Cloud Database Management Systems (DBMSs) in preference to the traditional DBMSs. Furthermore, the idea of having the data of multiple companies on the same database system

L. Garg et al. (Eds.): ISMS 2020, LNNS 303, pp. 46–57, 2022.
https://doi.org/10.1007/978-3-030-86223-7_5

is gaining popularity due to its benefits in terms of storage space and spreading the acquisition costs. Data is a vital asset and its quality depends on the proper administration and securing of the data held and processes allowed to manipulate the data. Consequently, data security must be ensured during the storage, processing and transmission stages [3]. Breaches in security undermine the overall consistency and availability of a system. Thus, appropriate security access control mechanisms should be in place to ensure that only authorized access is allowed. Nonetheless, although the importance of security is emphasized in literature, such a requirement is typically absent in the initial stages of development. Since security is a non-functional requirement, it is usually ignored or considered as a separate task towards the later parts of the lifecycle [4].

To tackle part of this problem, this paper proposes a framework which ensures that security in cloud-based database systems is encapsulated in the initial system design diagrams namely the Entity Relationship Diagrams (ERM), Entity Life History (ELH) and Data Flow Diagram (DFD). The aim of this framework is that of developing an automated process, able to implement security assertions prior to the storage of data online. Based on the requirements inputted and the system diagrams presented by the user, the framework can create specific security profiles, adorned with appropriate access control indications. Finally, following the implementation of access controls such as roles and privileges, this framework is capable of transforming the database diagrams enriched with the security features into SQL code to produce a secure database.

The remaining sections of the paper are as follows: Sect. 2 highlights the main literature works related to this paper, Sect. 3 highlights the methodology utilized and procedure followed for the implementation and testing of the proposed framework, whilst Sect. 4 points out at the overall achievements.

2 Literature Review

2.1 Cloud Computing and Security in Cloud-Based Systems

Cloud computing is currently one of the most prominent technologies and organizations are moving towards cloud-based systems due to this technology's numerous advantageous [1, 2]. According to a Forbes survey conducted in 2017 [5], 74% of all technological CEOs interviewed selected cloud computing as the technology having the greatest impact. The importance of the cloud is further emphasized by the fact that the development of cloud-based systems is currently 4.5 times faster than that of Information Technology itself and is expected to increase 6 times faster in 2020. Cloud computing is a computing model in which resources and services are accessed via an Internet connection. Its benefits include a reduction in the upfront investment of companies together with more predictable running costs. Moreover, capability planning for the provisioning of resources is no longer required from beforehand whilst data storage requirements can be quickly met through known elasticity features available [6]. These data processing requirements are supported by state-of-the-art DBMS which offer storage of huge amounts of collective data and leverage an acceptable and manageable data security. Regardless of these benefits some organizations are still weary in the adaption of cloud-based systems and this is primarily due to security concerns [3]. Security is vital in any system and even more so in cloud-based ones. This is because data would no longer be stored on-site but at a remote location which might possibly be shared [4].

Security related to databases utilized in cloud-based systems is an eclectic topic which many researchers have described and attempted to address [7]. Multiple models have been developed focusing upon security feature implementations. Singh et al. [8] presented a model which improves security features by utilizing the privacy preservation model approach, whilst Subashini et al. [9] examined security concerns in the area of cloud-computing by analyzing and evaluating existing literature. Although, it is known that security artefacts are introduced from the initial design stages, there is a research gap in work for the creation of general frameworks.

2.2 Security Implementation in Database Design

The implementation of security from the initial stages of database design is not only advantageous but indispensable [10]. Security attacks on database systems undermine the whole system and should hence be given due importance. An example of a security attack related to DBMS occurred in May 2019 whereby sixteen years of insurance holders' data corresponding to 1.3 billion records were leaked from the database of First American Financial Corporation [11]. Research to avoid such breaches have been conducted by many including Pourzargham [12] who identified two countermeasures used in the prevention of onsite database-specific attacks. Moreover, Basharat et al. [13] developed a semantic data model, able to extract constraints from an ERM.

Security is a vital facet which should be considered by all stakeholders. In traditional, onsite database systems security configurations are stored and enforced within the organization itself. In contrast, security requirements in cloud-based database systems are mainly enforced by the CSP because the overall security of the system is dependent on the weakest security feature. Nonetheless, enough flexibility should be offered to the tenants so that specific security conditions and features are implemented in conformance with their preferred access control mechanisms and other security considerations.

The idea of utilizing database design diagrams upon which security artefacts are entrenched is appealing. ERMs together with ELHs and DFDs provide a global overview of the whole system specific for each tenant. Thus, if security features are immediately implemented on such diagrams, a comprehensive approach is ensured. Furthermore, the choice of diagrams is critical as the right balance must be established when it comes to the amount of details presented. Providing too much detail would make security specifications too complicated, whilst providing too little details tends to gloss over the security requirements by making most data access clearance an all or nothing approach.

2.3 Security Access Control Mechanisms

One essential method of ensuring security in database system is by utilizing access control mechanisms. There are four main types of such mechanisms – Discretionary Access Control (DAC), Mandatory Access Control (MAC), Role-Based Access Control (RBAC) and Attribute-Based Access Control (ABAC) each of which is effective against a subset of security threats [14]. In DAC, those having administrative rights provide access rights to identifiable users. Provided that a subject has been granted a privilege on an object, he can then disseminate the privileges to other subjects. DAC is the most frequently implemented access control mechanism due to its flexibility. Nonetheless, it has some innate limitations including the risk of Trojan horse attacks [14, 15]. MAC is a mechanism which is less flexible than DAC whereby a security label is assigned to both the objects and the users. Users are then able to access objects according to this security label [15, 16] through a set of axioms and rules. RBAC is quite a novel mechanism, developed to be able to deal with modern technologies including cloud computing [17]. It can be easily integrated with DAC mechanisms by assigning users to roles. For example, 'user1' and 'user2', who are managers of different departments might both be assigned the role of 'manager'. This role would then define any access rights or privileges which are common to all managers in the company. The last form of access control mechanisms is ABAC whereby access rights are granted based on subject, resource and environmental attributes [17].

Irrespective of the access control mechanisms selected, it is vital that the maximum amount of security coverage is catered for. Furthermore, the design adopted needs to be flexible and at the same time manageable by the CSP or end users (i.e. when system is live). This would reassure the end users whilst enabling a CSP to offer better end products to their clients. Furthermore, if there are some known threats which are not addressed, then these should be clearly stated in the Service Level Agreement (SLA).

3 Methodology and Framework Development

The framework proposed in this paper consists of five main stages. Firstly, the system design diagrams (i.e. the ERM, ELH and DFD), representing a complete system are presented to the framework in JSON format whose constructs follow a set of rules. These are then analyzed through the construction of an entity-to-entity (adjacency) and entity-to-attribute (incidence) matrix to ensure completeness and reachability. Following such analysis, the third stage deals directly with access control mechanisms and is responsible for granting privileges on selected roles and database objects as required. Moreover, a Create, Read, Update, Delete (CRUD) matrix is used whenever a privilege is given to a role so that all required accesses to functions and objects are envisaged.

Based on the design data provided together with the security access control mechanisms outlined, the fourth stage of the framework involves the generation of a security report, responsible for providing overall security feedback on the database being constructed. Such feedback marks any issues related to reachability and isolation of database objects together with information related to security mechanisms. The final stage creates a secure database system depicted via SQL code which considers both the system design diagrams presented and additional security constraints required for proper access control mechanisms. The architecture of the framework explained is depicted graphically in Fig. 1 and is explained subsequently here.

This approach allows security assertions to be created automatically specifically for each tenant whilst avoiding manual modifications. In cloud-based multi-tenant systems, a robust system is required to ensure appropriate data isolation and implementation of specific security requirements for each tenant. The data set utilized for the testing of such a framework was 'DVD Rental' [18], since it is easily depicted and can be obtained online. The DBMS utilized for the testing of the result is PostgreSQL.

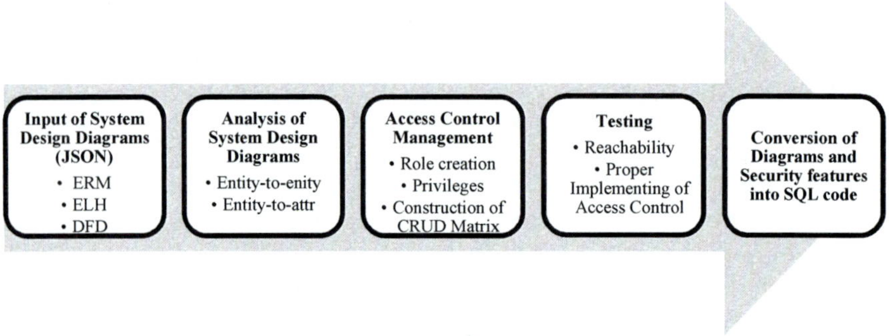

Fig. 1. Framework's architecture

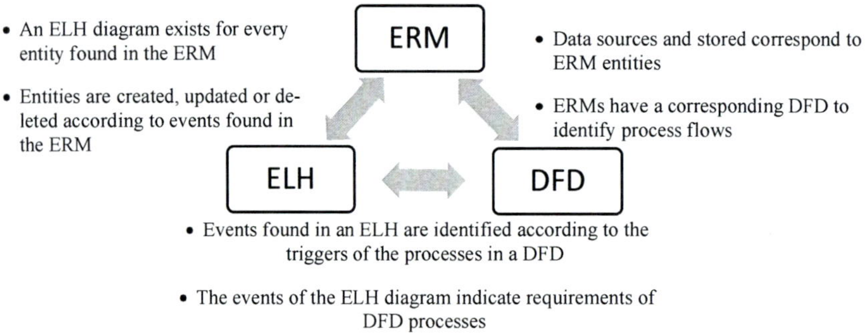

Fig. 2. Relationship between ERM, ELH and DFD

3.1 Inputting of System Design Diagrams

The first stage of the framework is that of receiving the system design diagrams so that these can then be further analyzed in the subsequent stages. The diagram displayed in Fig. 2 illustrates the relationship between each of the three diagrams. Access control rights such as read, write and execute and privileges to roles can be granted and revoked directly from these diagrams, presented to the framework in the form of JSON files following EBNF rules. The development of EBNF rules specifically for each file, define the grammar and syntax for each diagram whilst ensuring consistency throughout. There are 39 set of rules to encode an ERM and the encoding covers: strong and weak entities, n-ary relationships with cardinalities and participation constraints, single, multi-valued and composite attributes. The specifications related to the ERM starts off with the definition of the strong entities and their attributes. If required, these are then followed by definitions of weak entities present in the system and the relationships which exist between the entities. Each relationship is listed in the appropriate section and the name of the relationship together with the entities, cardinalities and participation is defined. Figure 3 illustrates three entities, 'Film', 'Language' and 'File_actor' with some of their attributes. It can be observed that the ERM encoding has a hierarchic structure.

There are 5 set of rules to encode an ELH and the encoding covers mainly the entities and list of processes for each entity. Each process might be divided into other processes,

```
{
"strong entities":[
{"strong entity name":"film", "attributes":[{"attribute name":"film_id",
"attribute type":"integer", "required":"yes",
"attribute mode":["non-derived","single","primary key simple"]}]},

{"strong entity name":"language", "attributes":[{"attribute name":"language_id",
"attribute type":"integer", "required":"yes",
"attribute mode":["non-derived","single","primary key simple"]}]},

{"strong entity name":"film_actor", "attributes":[{"attribute name":"actor_id",
"attribute type":"integer", "required":"yes",
"attribute mode":["non-derived","single","primary key simple"]}]},

"SE-SR-SE relationships":[
{"relationship name":"is","strong participating entity 1":
{"strong participating entity name":"language",
"participation in relationship":"partial","cardinality":"1"},
"strong participating entity 2":{"weak participating entity name":"film",
"participation in relationship":"total","cardinality":"N"}},

{"relationship name":"has","strong participating entity":
{"strong participating entity name":"film",
"participation in relationship":"partial","cardinality":"M"},
"weak participating entity":{"weak participating entity name":"film_actor",
"participation in relationship":"total","cardinality":"N"}}]
}
```

Fig. 3. Encoding of ERM

```
{
"entity":"Film","processes":[
{"processname":"Acquisition of new film","option":"none"},
{"processname":"Lifetime of film", "option":"none","processes":[
{"processname":"Update of film information details","option":"*"},
{"processname":"Availability of film changes", "option":"*","processes":[
{"processname":"Film returned", "option":"0"}]}]}]
}
```

Fig. 4. Encoding of ELH

each of which might be iterative or selective. The encoding starts off with the name of the entity, followed by all states in the ELH enclosed in brackets. The states are listed in accordance to their level whilst the '*' or 'o' after the state name indicates whether the state is iterative or selective. Figure 4 displays the encoding of the 'Film' ELH. It can be noted that six processes are associated with this entity and the processes are listed in a sequential manner.

There are 11 set of rules representing a DFD encoding which include entities, stores, processes and data flows listed in this order. The entities, stores, and processes are separated by commas and enclosed in square brackets. The data flows list the two entities involved in the transfer of data, preceded by its name. An example of a DFD depicting the relationship between the 'Customer' and 'Rental' tables is depicted in Fig. 5.

As already highlighted in Fig. 2, the three above-mentioned diagrams are interrelated. Every entity specified in the ERM should have a corresponding ELH representation

```
{
"external entities":[{"external entity name":"Customer"},{"external entity name":"Rental"}],
"data stores":[{"data store name":"FilmInformationLibrary"}],
"processes":[{"process name":"RentFilm","level":1},{"process name":"ReturnFilm","level":2}],
"dataflows":[{"dataflow name":"filmRequest",
"inward connected entity":"RentFilm","inward entity type":"process",
"outward connected entity":"Customer","outward entity type":"external entity"},
{"dataflow name":"bill",
"inward connected entity":"Customer","inward entity type":"external entity",
"outward connected entity":"RentFilm","outward entity type":"process"},
{"dataflow name":"rentalInformation",
"inward connected entity":"RentFilm","inward entity type":"process",
"outward connected entity":"Rental","outward entity type":"external entity"},
{"dataflow name":"filmInformation",
"inward connected entity":"FilmInformationLibrary","inward entity type":"data store",
"outward connected entity":"RentFilm","outward entity type":"process"},
{"dataflow name":"rentalInformation",
"inward connected entity":"ReturnFilm","inward entity type":"process",
"outward connected entity":"Rental","outward entity type":"external entity"},
{"dataflow name":"filmInformation",
"inward connected entity":"ReturnFilm","inward entity type":"process",
"outward connected entity":"FilmInformationLibrary","outward entity type":"data store"},
{"dataflow name":"filmInformation",
"inward connected entity":"FilmInformationLibrary","inward entity type":"data store",
"outward connected entity":"ReturnFilm","outward entity type":"process"},
{"dataflow name":"receipt",
"inward connected entity":"ReturnFilm","inward entity type":"process",
"outward connected entity":"Customer","outward entity type":"external entity"},
{"dataflow name":"information",
"inward connected entity":"Customer","inward entity type":"external entity",
"outward connected entity":"ReturnFilm","outward entity type":"process"}]
}
```

Fig. 5. Encoding of DFD

whilst each process specified in the ELH should be listed in the DFD. Furthermore, each data store specified in the DFD should have been defined as an entity in the ERM.

3.2 Analysis of Diagrams

The next stage involves the analyses of the three structured diagrams described in Sect. 3.1 to ensure their correctness in terms of reachability, prior to the implementation of any security procedures on such diagrams. This stage is divided into two main parts after reading in respective JSON file representing the structured diagrams:

Creation of Entity-to-Entity and Entity-to-Attribute Matrix
This step involves the computation of an entity-to-entity matrix. Such a matrix is a form of adjacency matrix whereby the relationship between all entities in the database is illustrated. Relationships might differ in cardinality and can be either one-to-one (1:1), one-to-many (1:N) or many-to-many (M:N). If the relationship is a ternary one, the keyword 'ternary' is displayed in the cell combining two of the entities and the remaining entity is listed. Figure 6 depicts an example of entity-to-entity matrix depicting the relationships between the tables encoded in the JSON file depicted in Fig. 3.

	Film	Language	Film_actor
Film		is 1:N	has M:N
Language	is 1:N		
Film_actor	has M:N		

Fig. 6. Entity-to-entity matrix for 'DVD rental' database

The generated matrix is then analyzed through queries to ensure that each entity is at least connected to one other entity. Such a connection would guarantee that no orphan entities are found in the database being created i.e. ensuring that all entities that are connected are indeed marked as so.

The other matrix which is created by the framework is a form of incidence matrix referred to as the entity-to-attribute matrix. Even this matrix is mostly based on information extracted from the encoded ERM. It displays all attributes of each entity found in the database. Figure 7 illustrates the part of the matrix appertaining to the 'Language' and 'Film_actor' table.

	Language_id	Name	Last_update	Actor_id	Film_id
Language	PK	Att	Att		
Film_actor			Att	PK	PK

Fig. 7. Entity-to-attribute matrix for 'Language' and 'Film_actor' tables

The attributes included in the entity-to-attribute matrix appertain to three main types of attribute context – part of a primary key, part of a foreign key and those appertaining to each individual entity.

3.3 Access Control Management

The Access Control Management part of the framework adorns the three system design diagrams mentioned in previous sections with data security instructions. To be able to capture most scenarios dealing with cloud-based system a mix of DAC and RBAC are utilized in this framework. Thus, the main objective of this stage is that of granting access rights and permissions on specific objects or processes to roles. This method is quite a rigorous task since a good overview of each of the roles' tasks should be identified to determine the privileges required by each role. Furthermore, it should be ensured that privileges are granted on the concept of least privilege so that the maximum amount of security is guaranteed. This is determined by considering the DFD to determine which are the objects required for each function to be executed. The context of this framework for the implementation of security features with respect to access control is restricted to create, execute, read, write and delete privileges for roles on artefacts/entities or functions. Implementing such security controls directly on the system design diagrams would assist the database administrator and ensure more clarity even for customization when system is live.

Creation of Roles

Since discretionary and role-based access controls are being utilized, all roles required for the proper execution of the system must be created. The list of all roles is made available to the framework. Following the creation of such roles, privileges can then be granted on objects to each role.

Granting Privileges and Access Rights on Base Tables

The implementation of security privileges on base tables (e.g. an implementation of an entity) involves the granting of Select, Insert, Update and Delete privileges. These privileges are granted to roles directly from the ERM in accordance with the data required by each role to be able to perform their assigned task.

The framework then checks the privileges file containing to ensure that such a privilege has not already been granted. Provided that is not the case the privilege is granted.

Granting Privileges and Access Rights on Functions

The granting of privileges on to a function is a much more complex task. This is because the utilization of all three system design diagrams is required. For a role to be granted execute privilege on functions, the role should have first been granted privileges on the objects/base tables required by the function.

The DFD is essential in the identification of such privileges. Data flows going out from a data store and into a process represent a SELECT privilege on the data store. Data flows going into a process represent a SELECT privilege on base tables whilst those going out of a process into data stores represent INSERT, UPDATE or DELETE privileges on the data store. To identify which privilege is required, the ELH diagram is then utilized. The first sub-tree of the ELH represents the initial state (i.e. birth) of an entity since processes are depicted in an ELH in a sequential manner. Thus, if an EXECUTE privilege is required on a function which is at the beginning of the ELH sub-tree, an INSERT privilege would be required. Following the same line thought, the sub-tree in the middle represents any modifications which might, hence corresponding to and UPDATE privilege. Lastly functions corresponding to the end of an ELH sub-tree correspond to a DELETE privilege since these would represent the death of the entity.

Once it is determined which kind of EXECUTE privilege is to be assigned, there is another check which needs to be conducted prior to the granting of the said privilege. The framework must ensure that a role has access to all of the objects required by the function. This can be established by determining which base tables are required from the DFD and the ERM and checking the privileges which have been assigned. Once such a check is verified, the required privilege can be granted to the role on the function.

CRUD Matrix

All privileges assigned to each of the roles found in the database are stored in a CRUD Matrix. Since checks are conducted prior to the granting of privileges to ensure that such a privilege has not already been granted, this matrix is extremely useful.

3.4 Testing

The testing of the proposed framework ensures reachability and proper implementation of access controls. Multiple test cases associated with system design diagrams were implemented, as was the case in research works such as [18]. The testing conducted throughout the development of this framework amalgamates such test cases and are explained subsequently.

Testing for Completion and Consistency
One of the tests which is conducted in this framework is verifying that all objects found in the database are necessary and reachable. This verification task is divided into two main parts. Firstly, the entity-to-entity matrix is examined to ensure that each entity is associated to at least one other entity thus, ensuring that all items in the ERM (i.e. base tables) are reachable. The items found in the ELH and DFD can then be checked for reachability by examining the list of roles and privileges. All roles in the database are examined to check which privilege is applied to each role. Provided that a role doesn't have at least one privilege associated with it, the role is either not required and can thus, be deleted or else there is a flaw in the access controls assigned. Furthermore, for a user to be able to access or alter the database state he must be able to execute at least one function. Thus, there is surely a problem in the security of the whole database structure if a role has been granted access on base tables but not on any function.

Testing for Vertical Coverage of Attributes
Another test which is conducted to ensure the correct functioning of the framework is vertical coverage testing. Vertical coverage ascertains that all attributes in all tables are required. This is ensured by guaranteeing that every attribute has at least one role which has been granted select, insert, delete or update privilege on it.

If one of the attributes is not covered, it would either mean that that attribute is not required or else there is a flaw in the security implementation which needs to be addressed so that the correct role is able to access the attribute.

Testing for Correct Access Controls in Terms of Execute Privileges on Functions
The last test conducted, takes place when an execute privilege is granted to a role on a function. It must be ensured that a role has enough privileges to be able to execute that function. This entails having access to any objects accessed or modified by the function and ensuring that privileges are being applied on the least privilege concept.

3.5 Creation of Secure Database (Conversion of Secure Diagrams to SQL Code)

The final stage of the framework is related to the creation of a secure database. Once the initial system design diagrams are embellished with the required security features specified throughout Sect. 4, the diagrams, matrixes and roles are converted into SQL code so that a database is constructed that reflects the security instructions.

This framework is a general-purpose one, in the sense that once the result diagrams and security features are satisfactory, they can be converted into any DBMS software. The cloud-based DBMS software used for this case study was PostgreSQL.

Furthermore, the framework is easily reusable in the sense that should any underlying specifications and assumptions change during run-time, the system design diagrams, and access controls can be easily re-visited to cater for this change.

4 Conclusions

The framework proposed in this paper, supports the development of secure cloud-based database systems. It allows the incorporation of security features from the initial stages of system development by endorsing security features on system design diagrams.

Although numerous research works highlighted the importance of security considerations from the initial design stages, most of them fell short in developing frameworks based on such an indications. Some research works such as that proposed in [19] implement security requirements on a singular design diagram. Although this is definitely a leap-forward, a single diagram is not able to depict a comprehensive view of the system. Thus, additional features have to be kludged in to try and replace what is missing from the representation of the diagram. The utilization of three interrelated but complimentary system design diagrams to entrench security profiles presented in this paper is a novel one. A complete system representation is presented to the proposed framework via an ERM, ELH and DFD diagrams. The analysis of these three interrelated diagrams ensure that mistakes in the security design are avoided. Security assurance is then further augmented in this framework, through the ability of creating specific security profiles for each tenant based on various implementation of access rights in conformance with individual tenant's requirements.

An advantage of such a framework is that it is re-usable in the sense that should any new requirements be identified in the security regime of the system; the database administrator can re-use the framework and make the required alterations to both the design diagrams and privileges. Even though, discretionary and role-based access controls, considered in this framework cover the majority of security concerns in cloud-based database systems, it would be beneficial if such a framework is extended so that MAC is incorporated too. Nonetheless, the creation of such a framework is a step forward in the area of cloud-based security since it ensures that security features are considered from the very initial stages of software development life cycle.

References

1. Catteddu, D.: Cloud computing: benefits, risks and recommendations for information security. In: Serrão, C., Díaz, V.A., Cerullo, F. (eds.) IBWAS 2009. CCIS, vol. 72, pp. 17–17. Springer, Heidelberg (2010). https://doi.org/10.1007/978-3-642-16120-9_9
2. Dubey, S.K.: Security and privacy in cloud computing. Int. J. Comput. Sci. Mob. Comput. **2**, 123–125 (2013)
3. Januzaj, Y., Ajdari, J., Selimi, B.: DBMS as a cloud service: advantages and disadvantages. Procedia Soc. Behav. Sci. **195**, 1851–1859 (2015)
4. Duraisamy, G., Ghani, A., Zulzalil, H., Abdullah, A.: Analysis of access control model for data Security and Privacy on Multi-Tenant SaaS. Adv. Sci. Lett. **24**, 1619–1622 (2018)
5. Columbus, L.: Roundup of Cloud Computing Forecasts, 2017. Forbes Report (2017)

6. Zhang, Q., Cheng, L., Boutaba, R.: Cloud computing: state-of-the-art and research challenges. J. Internet Serv. Appl. **1**(1), 7–18 (2010). https://doi.org/10.1007/s13174-010-0007-6
7. Aldossary, S., Allen, W.: Data security, privacy, availability and integrity in cloud computing: issues and current solutions. Int. J. Adv. Comput. Sci. Appl. **7**, 485–498 (2016)
8. Singh, A., Sharath, K.J., Muralidhar, A., Muralidhar, A.: A secure multi-tenant model for SaaS system. Int. J. Comput. Appl. **90**(15), 1–5 (2014). https://doi.org/10.5120/15793-4468
9. Subashini, S., Kavitha, V.: A survey on security issues in service delivery models of cloud computing. J. Netw. Comput. Appl. **34**, 1–11 (2011)
10. Matt, B.: Introduction to Computer Security. Pearson Education India (2006)
11. Irwin, L.: List of data breaches and cyber-attacks in May 2019 – 1.39 billion records leaked (2019). https://www.itgovernance.co.uk/blog/list-of-data-breaches-and-cyber-attacks-in-may-2019-1-39-billion-records-leaked
12. Pourzargham, H.: Importance of security in database. Int. J. Comput. Sci. Netw. Secur. **15**, 29–31 (2015)
13. Basharat, I., Azam, F., Muzaffar, A.W.: Database security and encryption: a survey study. Int. J. Comput. Appl. **47**, 28–34 (2012)
14. Hasani, S.M., Modiri, N.: Criteria specifications for the comparison and evaluation of access control models. Int. J. Comput. Netw. Inf. Secur. **5**, 19 (2013)
15. Meghanathan, N.: Review of access control models for cloud computing. Comput. Sci. Inf. Sci. **3**, 77–85 (2013)
16. Aluvalu, R., Muddana, L.: A survey on access control models in cloud computing. In: Satapathy, S.C., Govardhan, A., Srujan Raju, K., Mandal, J.K. (eds.) Emerging ICT for Bridging the Future - Proceedings of the 49th Annual Convention of the Computer Society of India (CSI) Volume 1, pp. 653–664. Springer, Cham (2015). https://doi.org/10.1007/978-3-319-13728-5_73
17. Lo, N.W., Yang, T.C., Guo, M.H.: An attribute-role based access control mechanism for multi-tenancy cloud environment. Wirel. Pers. Commun. **84**(3), 2119–2134 (2015)
18. PostgreSQL. PostgreSQL Sample Database. http://www.postgresqltutorial.com/postgresql-sample-database/. Accessed 20 Sept 2020
19. Giuri, L., Iglio, L.: A role-based secure database design tool. In: Proceedings of the 12th Annual Computer Security Applications Conference, pp. 203–212. IEEE (1996)

Interactive Workbook for Effective Virtual Laboratories

S. Krishna Bhat, Shemin Anto, Narayanan V. Eswar, Shreyas S. Kumar, and G. Pankaj Kumar(✉)

Department of Computer Science and Engineering, Federal Institute of Science and Technology (FISAT), TM Hormis Nagar, Mookkannoor P O, Angamaly, Ernakulam, Kerala, India

Abstract. The COVID-19 pandemic has forced academic institutions to switch from traditional teaching-learning to fully digital mode. The traditional programming lab sessions are replaced by video lectures, notes, and assignment submission through LMS. Manual grading and debugging of the program results in delayed feedback. The existing auto-graders are designed to check the programs' correctness, and they cannot enhance learning. The interactive workbooks we propose are similar to the popular Jupyter notebooks but oriented more towards enhancing the teaching-learning process and providing immediate feedback. The survey results showed that 70% of the students believed that interactive workbooks enabled them to understand the problem and made them capable of solving it in incremental steps. 65% of the instructors added that interactive workbooks could supplement the physical lab sessions' teaching-learning process.

Keywords: LMS · Virtual labs · Notebooks

1 Introduction

The current pandemic situation, due to COVID-19, has brought up many changes worldwide. Though impermanently, almost every sector of human life has been affected to a great extent. Education being one among many, had to shift from the traditional way to digital modes. Teachers and students have been forced to depend only on e-learning. Though the learning process is being carried out with LMS's help, issues arise when students have to deal with practical sections. In contrast to the general scenario where students perform coding tasks in labs and receive feedback about their work from faculties immediately, at present, they are made to follow lecture videos and do coding assignments, and submit the same to the teacher through LMS. Popular platforms such as Jupyter Notebook are widely used to provide an environment for program development and facility to share learning resources [9]. It supports theoretical and practical classes as an aid for the enhancement of the learning and practical skills of students.

L. Garg et al. (Eds.): ISMS 2020, LNNS 303, pp. 58–66, 2022.
https://doi.org/10.1007/978-3-030-86223-7_6

The current python notebooks are efficient tools for demonstrating and explaining data science and machine learning programs. However, a drawback is that most learners will understand the code and will never gain the capability to solve the problem by themselves. This paper describes an overview of the interactive workbook we have developed.

Our interactive workbook may be viewed as an alternative to notebooks like Jupyter Notebook, with the additional capability of supporting interactive learning. It facilitates incremental learning methodology where students are made to reach the final goal step by step, performing simpler tasks. The whole program is divided into sub-blocks, and students have to complete each block independently, forming a solution to the complex problem. It helps them to understand the logical concepts clearly. Rather than merely providing the solution for a problem, this workbook focuses on enabling students to form solutions of their own by leading them to provide instructions, feedback, and hints.

The main intention of such a workbook is to reduce the instructor's work, where they only need to focus on the main topic of the lab session and need not get distracted by technical aspects of the program itself. Thus, it encourages the instructor to increase the quality of their lectures and inculcates an attitude of self-dependence and exploration. This can improve lab sessions' effectiveness and improve the students' programming logic without any direct involvement from the instructor.

2 Related Works

Hamza Manzoor et al. proposed an auto-grading tool for Jupyter notebooks. The proposed tool enables the delivery of immediate feedback to the students. The survey results indicated that 80% of students agreed that immediate feedback could improve performance [1].

H Hermansyah et al. proposed using guided inquiry models on virtual machines to improve students' understanding of heat concepts. The paper suggests that learning can be improved using a discovery-based model where students formulate and evaluate the hypothesis for themselves [2].

DIANA BOGUSEVSCHI et al. proposed a 3d virtual environment for teaching and studying physics. The proposal enables an immersive education by providing a 3d simulation of real-life and creating a sensory experience [3].

Zuhoor Al-Khanjari et al. proposes a moodle based virtual lab for computer science students. The methodology incorporates a virtual lab with a soft tool integrated into an LMS environment allowing accessibility from anywhere at any time [4].

Thomas Staubitz et al. proposes a method for automating the evaluation of massive open online courses. The paper helps in the correction of assignments of a large number of students who participate in MOOCs and reduces the tutor's workload [5].

Yu-Chang Hsu et al. talks about the educators' experience in online graduate courses on mobile app development. The paper discusses educators with limited programming skills learning to design a mobile app with peers and instructor guidance [6].

Jukka Koskinen et al. proposes a user-friendly online walkthrough approach on task level programming. The paper proposed using a tablet as the user interface to acquire paths using an external F/T sensor and making a program to use this path for machining tasks [7].

Martinha Piteira et al. is a literary review on gamification of online courses for learning computer programs. The paper also provides a theoretical framework to help guide teachers in gamification of online courses [8].

3 Methodology

The interactive workbooks are designed to provide continuous interaction with the students using audio, video, text, and feedback. The workbook divides each program into logical segments called sections based on the indications provided by the instructor. The interactive workbook is currently supporting Python; this may be extended to support other languages.

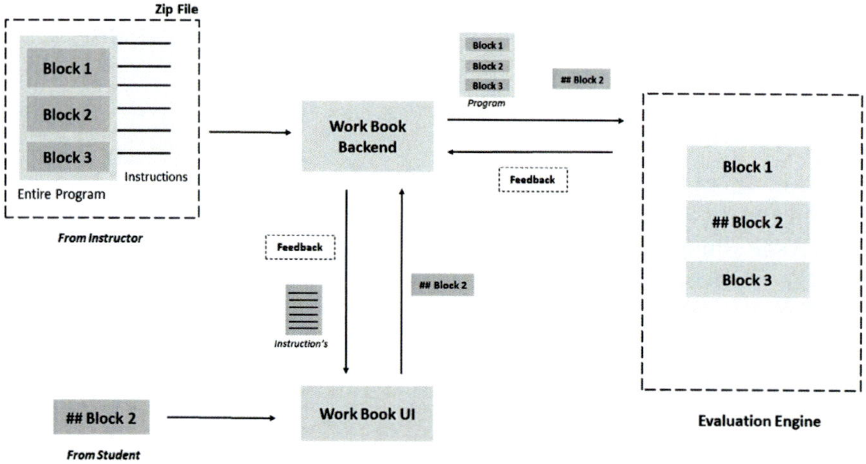

Fig. 1. Interactive workbook architecture.

3.1 Section

A section is a logical division of the program for which the workbook is designed. The faculty has to divide the problem into multiple sections. Each section contains an instruction block, code block, and feedback block.

3.2 Instructor's Role

The workbook aims at minimizing the role of the instructor to encourage a self-learning attitude in the students. The instructor is required to divide the program into multiple sections and provide the instructions and restricted code blocks along with the test cases for each section via HTML and other media. The workbook is saved as an XML file. The instruction should upload the compressed workbook along with associated resources to the workbook server. The workbook uses these files to set up the environment for the students to learn.

3.3 Student's Role

The student workbook will contain instructions and an empty code segment. The student will be provided immediate feedback upon submission, and the student can use the feedback area for additional hints.

3.4 Instruction Block

The faculty can upload instructions, audio, and video in the instruction area to explain the section's objective. The instructor needs to explicitly provide the names of the variables that the user should avoid/use. The variables may be used in the upcoming regions. The objective of the section may be the computation of intermediate results or the development of a function. The instruction block will be loaded in the student view in read-only mode. Figure 2 shows how the instruction and code blocks looks like.

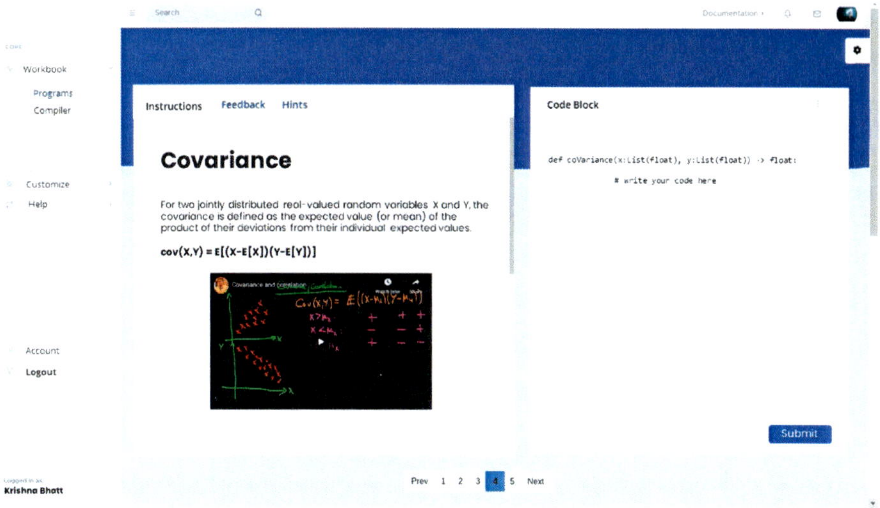

Fig. 2. Instruction block and code block.

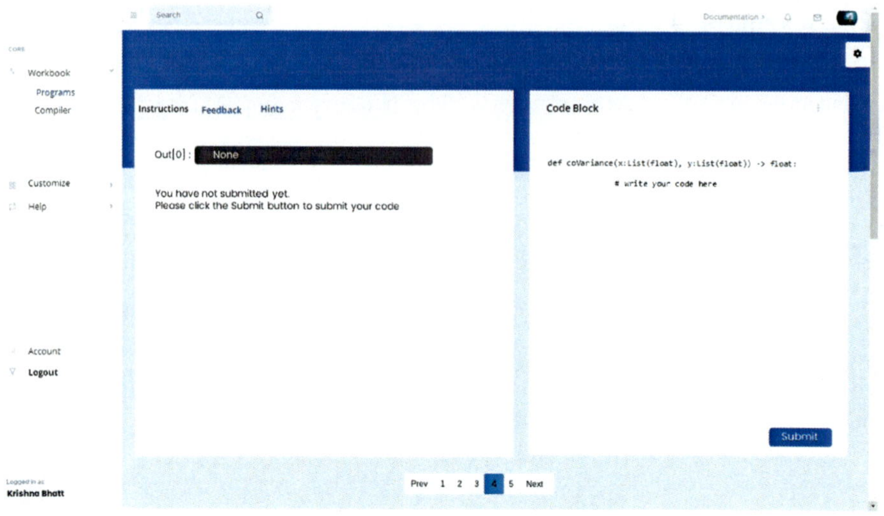

Fig. 3. Feedback block and output.

3.5 Code Block

The workbook contains two types of code blocks restricted code blocks and assignment code blocks. The instructor should provide working code in assignment block and restricted blocks to initialize variables and display the variables. The restricted code blocks can be set as read-only or hidden in the student view. The students can write their code in the assignment code block. The student can click the submit button to submit the code. The code evaluation engine will evaluate the code, and the output will be displayed in the output area.

3.6 Feedback Block

The feedback to the user is provided in the feedback block along with the output. In case of errors, the students can update their code and use the hint section to understand the problem better. A preview can be seen in Fig. 3. The hints may be provided on multiple levels. The instructor can upload hints regarding the general strategies, common issues, explanation of library routines, etc. in the feedback area.

3.7 Code Evaluation Engine

The code evaluation engine evaluates the code snippet from the assignment code block by replacing the instructor's code snippet. The instructor can provide a test set in the restricted code blocks. The feedback will contain information about the test cases in which the student code failed. The code evaluation engine inserts predefined patterns before the print statement of interest and extracts the output to be examined from the standard output. Fuzzy matching is performed on the output to deal with variations in the student output. This is done to ensure that the student's standard output does not interfere with the required output to pass the test cases.

3.8 Workbook Backend

The backend server manages the workbooks for various exercises and programs, serves the UI, and also manages the Code Evaluation engine. It keeps track of the current code block being attempted by the student and provides the corresponding restricted block to the engine, and the student's attempted code block to be evaluated by the engine. The server also extracts the required output from the total output evaluated by the engine, which is marked off by the engine with a predefined pattern. The server then verifies the test cases, and the feedback is provided to the student based on their performance.

4 Evaluation and Discussion

The workbook was used as part of internship, and the feedback survey was obtained from 40 students. A questionnaire containing ten questions were shared with students. We have chosen three out of these to be the most relevant, and have collected their results as follows. The students were asked whether the interactive workbooks were useful in bridging the gap between understanding the concepts and practical applications. Seventy percent of students agreed to question Fig. 4. The students were asked whether the interactive workbooks can be used as supplementary materials for regular labs. Sixty-five percent of students agreed to it Fig. 5. The students were asked whether the interactive workbooks can be used to replace the existing labs. Around 62% of students disagreed with it Fig. 6.

In their feedback, students added that during regular labs, the frequent mistakes are explained by the instructor. However, the interactive workbook currently does not contain explanations/details regarding common mistakes. In the regular lab sessions, faculties will give an overview of multiple ways in which the problem can be addressed. However, such insights are lost during interactive sessions. However, the majority of the students agreed that interactive workbooks are more efficient than the current methods used for digitizing the lab.

The instructors/TA were also provided a questionnaire containing ten questions. The instructors all gave a positive response towards the implementation of the interactive workbooks during the semesters. The feedback indicated that the process of creating the experiment is not easy, mainly due to the presence of multiple sections. However, the instructors agreed that interactive workbooks are useful in addressing a large number of students online. One instructor indicated that logically splitting the program into sections might be confusing to some instructors.

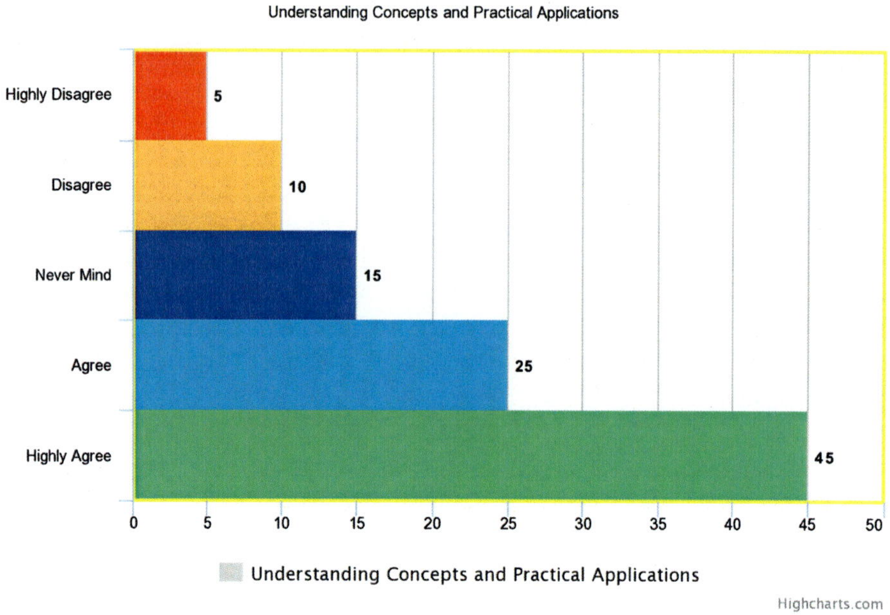

Fig. 4. Results of Q1: understanding concepts and practical applications

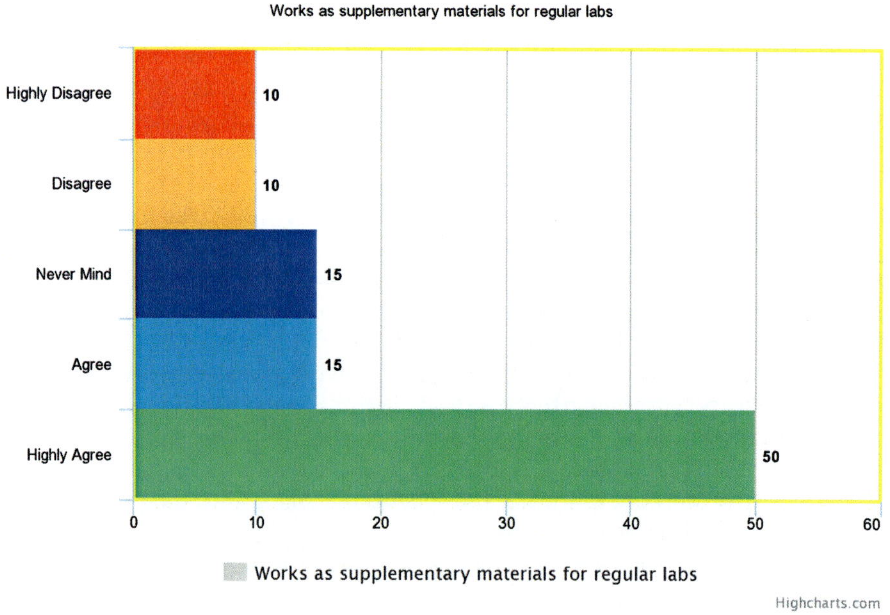

Fig. 5. Results of Q2: works as supplementary materials for regular labs

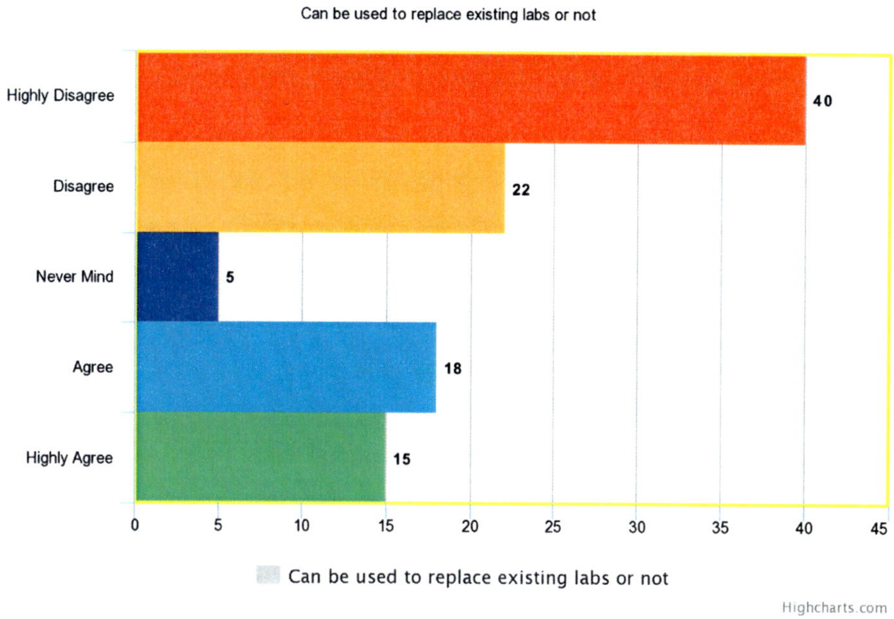

Fig. 6. Results of Q3: can be used to replace existing labs or not

5 Conclusion and Future Work

New innovative methods to enhance e-learning is in high demand as per the latest studies. The proposed Interactive Workbooks provide an efficient solution for digitizing the traditional programming labs. The process of breaking the problem into multiple segments and solving one segment at a time enables a student to understand the problem easier. The data from the workbooks may be used to obtain statistics regarding

- What portion of students are spending more time?
- What is the average time spend on each segment?

References

1. Hamza Manzoor, F., Amit Naik, S.: Auto-grading Jupyter notebooks. In: Proceedings of the 51st ACM Technical Symposium on Computer Science Education (SIGCSE 2020), pp. 1139–1144. Association for Computing Machinery, New York (2020). https://doi.org/10.1145/3328778.3366947
2. Guzmán, J.L., Joseph, B.: Web-based virtual lab for learning design, operation, control, and optimization of an anaerobic digestion process. J. Sci. Educ. Technol. **30**(3), 319–330 (2020). https://doi.org/10.1007/s10956-020-09860-6

3. Bogusevschi, D., Muntean, C. Muntean, G.M.: Teaching and learning physics using 3D virtual learning environment: a case study of combined virtual reality and virtual laboratory in secondary school. J. Comput. Math. Sci. Teach. **39**(1), 5–18. Waynesville, NC USA: Association for the Advancement of Computing in Education (AACE), 24 September 2020. https://www.learntechlib.org/primary/p/210965/
4. Al-Khanjari, Z. , Al-Roshdi, Y.: Developing virtual lab to support the computer science education in moodle. In: Proceedings of 2015 12th International Conference on Remote Engineering and Virtual Instrumentation (REV), Bangkok, pp. 186–191 (2015). https://doi.org/10.1109/REV.2015.7087290
5. Staubitz, T., Klement, H., Renz, J., Teusner, R. Meinel, C.: Towards practical programming exercises and automated assessment in massive open online courses. In: 2015 IEEE International Conference on Teaching, Assessment, and Learning for Engineering (TALE), pp. 23–30. IEEE, December 2015. https://doi.org/10.1109/TALE.2015.7386010
6. Hsu, Y.-C., Ching, Y.-H.: Mobile app design for teaching and learning: educators' experiences in an online graduate course. Int. Rev. Res. Open Distrib. Learn. **14**(4), 117–139 (2013). https://doi.org/10.19173/irrodl.v14i4.1542
7. Brunete, A., et al.: User-friendly task level programming based on an online walk-through teaching approach. Ind. Robot **43**(2), 153–163 (2016). https://doi.org/10.1108/IR-05-2015-0103
8. Piteira, M., Costa, C.J., Aparicio, M.: Computer programming learning: how to apply gamification on online courses? J. Inf. Syst. Eng. Manage. **3**(2), 11 (2018)
9. Cardoso, A., Leitão, J., Teixeira, C.: Using the Jupyter notebook as a tool to support the teaching and learning processes in engineering courses. In: Auer, M., Tsiatsos, T. (eds.) The Challenges of the Digital Transformation in Education. ICL 2018. Advances in Intelligent Systems and Computing, vol. 917. Springer, Cham (2019). https://doi.org/10.1007/978-3-030-11935-522

Detecting Presence of Masks and Violation of Social Distancing

Dhruv Bansal[(✉)] and Princy Randhawa

Manipal University Jaipur, Jaipur, India
dhruv.179403018@muj.maniapl.edu

Abstract.
Background: 2020 brought an epidemic with it which already have taken a lot of live all around the globe. The vaccine of this disease named COVID-19 is still under trails but there are some precautionary measures issued by WHO for people to keep themselves safe from getting infected and transfer it forward. However, some people are not taking this seriously and due to that government has also issued some rules and regulations for people who are not following the precautionary measures.

Methodology: In this study, we worked on how to find defaulters in real time world and identifying them between the group of number of people and ask them to follow the precautionary measures and if necessary, take actions against them. A convolutional neural network model was created and was implemented in finding the defaulters from a group of people.

Results: The model worked quite well in identifying people with and without face masks and also whether the people are following social distancing or not simultaneously through the webcam. The model was able to detect multiple number of people at once and also calculate distance between them and checking whether they are wearing masks or not.

Practical Applications: The research can be used in various places like factories, shops, roads, and other public places. In the areas where a number of people are working together, during this time it has become a necessity to have precautions. But due to some people's negligence it has become necessary for government bodies to keep check on these people and take necessary actions required and ensure the safety of others and also them. Therefore, it can used in surveillance system in the entire cities to keep check.

1 Background

Coronavirus hit India in around March and its effects has been increasing since then. The infected people the death toll has been kept on increasing since the day. This virus spreads through the tiny droplets that are secreted when we either sneeze or cough. These droplets either settle down on the virus and then pass onto the other person when the person touches that surface before the virus dies and then touch one's eyes or mouth

L. Garg et al. (Eds.): ISMS 2020, LNNS 303, pp. 67–73, 2022.
https://doi.org/10.1007/978-3-030-86223-7_7

or the places which gives passage to virus to enter into the body with the contaminated hands (Fig. 1).

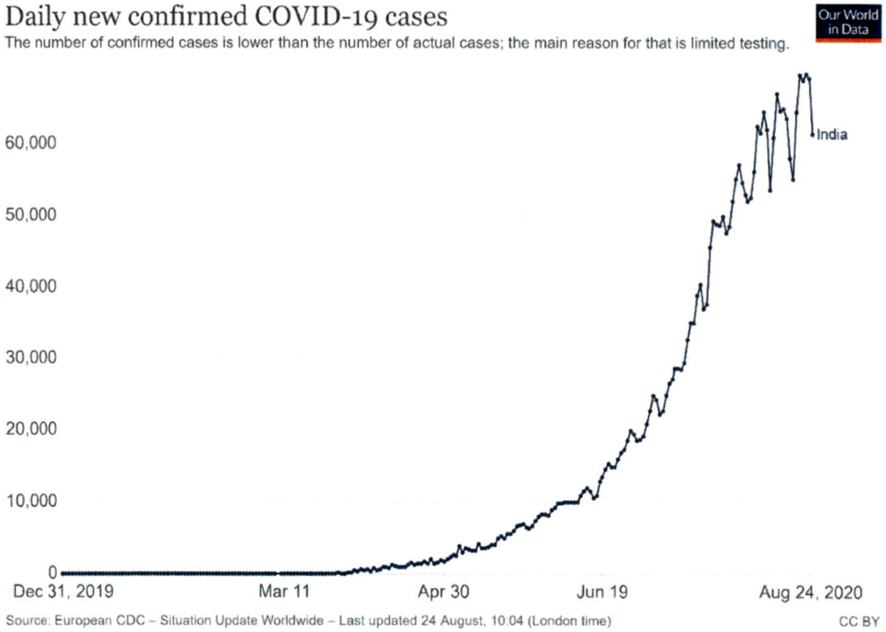

Fig. 1. Number of COVID-19 cases https://ourworldindata.org/coronavirus/country/india?country=~IND

As the number of cases kept on increasing government tried different methods to stop it from spreading but the only solid conclusion, they come up to be the precautions. Therefore, government issued rules for the people to take which were: Wearing face masks whenever leaving home, strictly follow social distancing i.e., keeping a minimum of 2 m distance as the droplets which are secreted while talking, coughing and sneezing are quite heavy that they settle down on ground before reaching to other person. The best results for keeping oneself safe from coronavirus were using mask. As the mask prevents those droplets from getting out in the air reducing risks to 70% (Fact-check: How much do masks reduce coronavirus 'contagion probability'?). Some people of the society are very careless about others as well as their own safety. They are not ready to follow the precautions the government have asked people to follow risking lives of everyone around them and theirs too. Due to these people it has forced government to keep check on these types of people and have issued some strict actions how are not following the rules.

In this research, we have introduced a model of detecting face mask and practicing of social distancing in real time both together rather than detecting one at a time. If we talk about the network architecture uses multiple models pre-defined and custom combined together for a better and robust model for face mask detection. Also, it has been quite difficult to find the data for social distancing, therefore, we used YOLO for detecting people in the real time video feed used for testing and the measure the distance between them. About the dataset, we used approx. 700 images of each category, i.e., masked, and non-masked people.

Further paper is divided into different sections. Section 2 comprises of the different method used for the same. The methodology has been explained in the Sect. 3. Whereas Sect. 4 and Sect. 5 comprises of the results and future work, respectively.

2 Related Work

2.1 Object Detection

Object detection is not a single step process but comprises of multiple level of image processing and extracting data from images. A well-known detector is the Viola-Joins detector, which is able to achieve real-time detection (Viola and Jones 2001). Features are extracted from the data using different methods like Haar-wavelet. Other algorithms are also there for feature extraction like SURF, ORB etc. These features are then used and are compared from the test data received from different sources and match the accuracy to give an output.

2.2 Convolution Neural Network

Rather than using handcrafted features, deep learning-based detector demonstrated excellent performance recently, due to its robustness and high feature extraction capability (Zou et al. 2019). CNN works in different ways depending on the type of the data. In this case we are using images as the data therefore, the data is passed through the CNN which extract features from the data or in layman's language the model learns from the data and some weights are generated which are used by the model to predict output over the data fed for testing or the real time data fed to the model.

3 Methodology

The network I used is not just a simple neural network but a mix of predefined models and custom models which are combined together for better results. I used starting layers of MobileNetV2, and the later layers were custom for face mask prediction. MobileNetV2 improves the performance of mobile models on multiple tasks and benchmarks as well as across a spectrum of different model sizes. (Sandler et al. 2018). Another reason to choose the MobileNetV2 was that it uses lightweight convolutions for feature filtering which help us to implement a deep and robust convolutional neural network without making it heavy to use on light GPUs. The later layers are custom because the output we want have only two categories whereas MobileNetV2 is for many more. The head of the model was the MobileNetV2 whose outputs were then used as the input for the base of the model. The base of the model had a max pooling 2D layer which was followed by flattening layer then a dense layer. After which a dropout layer was introduced which was followed by another dense layer which gave the predictions (Fig. 2).

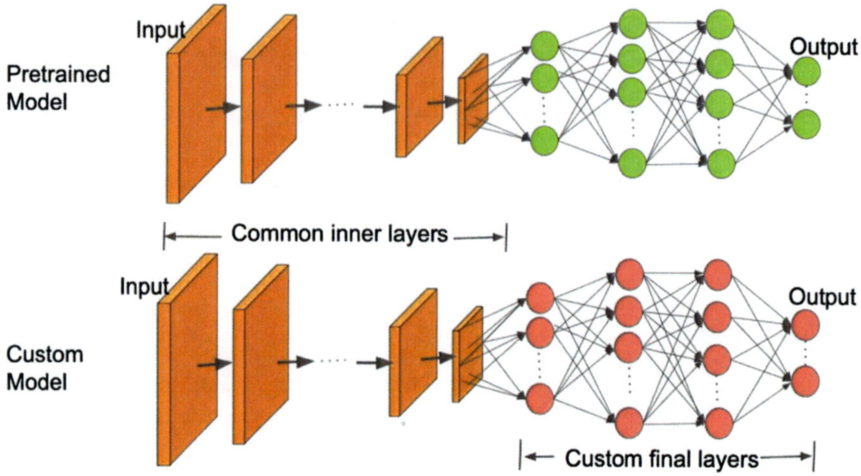

Fig. 2. Representation of the model https://www.learnopencv.com/image-classification-using-transfer-learning-in-pytorch/

The face are being detected using Caffe (Convolutional Architecture for Fast Feature Embedding) model. I have used Caffe model because of its speed and its ability to switch between CPU and GPU by setting a single flag to train on a GPU machine then deploy to commodity clusters or mobile devices (Jia et al. 2014). After the faces being detected those faces are being used as an input for the face mask detector model for detection of face masks. A threshold has been set for proper detection of face masks in real time. If the confidence is less than that of the threshold a red bounding box is created otherwise a green one. Simultaneously people are detected.

For the detection of people for social distancing we used specific class from YOLO. A single convolutional network simultaneously predicts multiple bounding boxes and class probabilities for those boxes; YOLO trains on full images and directly optimizes detection performance (Redmon et al., n.d.). The blobs created around the detections (people) were passed on through the algorithm for calculation of Euclidean distance between them and then comparing it to the threshold decided for determining whether the people are at safe distance or not simultaneously with detection of face masks on people faces. The threshold value of the distance is being mentioned in the form of pixels between two bounding boxes and the threshold value is chosen by multiple experiments practically and the chosen represents the distance of about 1.5 feet when measured in real time.

For the Euclidean distance, the centroids of two bounding boxes which are chose whose distance is to be compared are being calculated and then the those are used for calculating the Euclidean distance using the Eq. (1).

$$d(a, b) = \sqrt{}((bx - ax)^2 + (by - ay)^2 \quad (1)$$

where a = (ax, ay) is the centroid of one image and b = (bx, by) is the centroid of another and d(a,b) represents the Euclidean distance (CHAPTER 4 4. *METHODS FOR MEASURING DISTANCE IN IMAGES 4*, n.d.).

4 Results

Outstanding results were obtained from the neural network introduced. With almost 100% training and validation score. And great detection in real time feed from the cameras used to test the model. Below is the values we get after the model evaluation (Table 1).

Table1. Evaluation of the network

	Precision	Recall	F1-score	Support
With mask	1.00	1.00	1.00	138
Without mask	1.00	1.00	1.00	138
Accuracy			1.00	276
Macro avg.	1.00	1.00	1.00	276
Weighted avg.	1.00	1.00	1.00	276

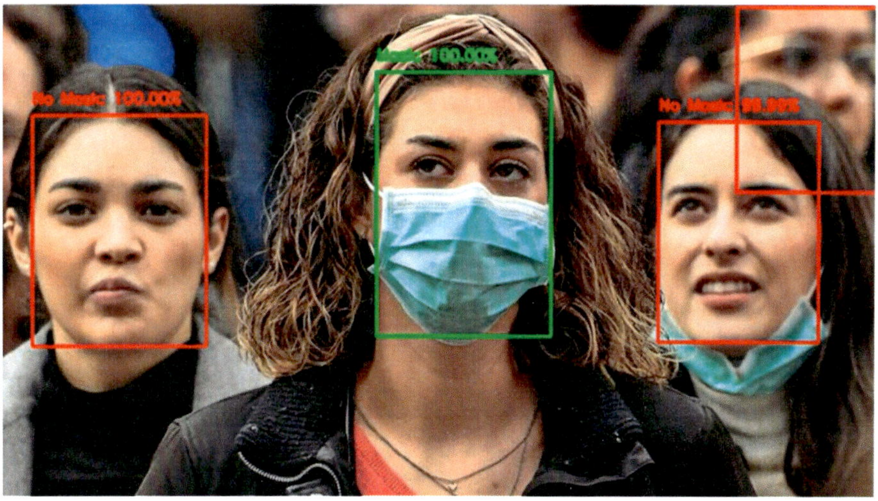

Fig. 3. Output of face mask detector model

Figure 3 is the output we got after using the picture as an input for our model. As seen in the picture the people who are without masks are bounded by a red box and people with mask are bounded by green box with a "No Mask: (percentage)" label and "Mask: (percentage)" label respectively above the box where the percentage is the confidence level of our model about its prediction.

Figure 4(a) and Fig. 4(b) are the frames taken from the output of the live video feed which is being used as an input in our final model. The detection and output given is done at the very moment. Figure 4(a) is the point where the subjects are at optimum

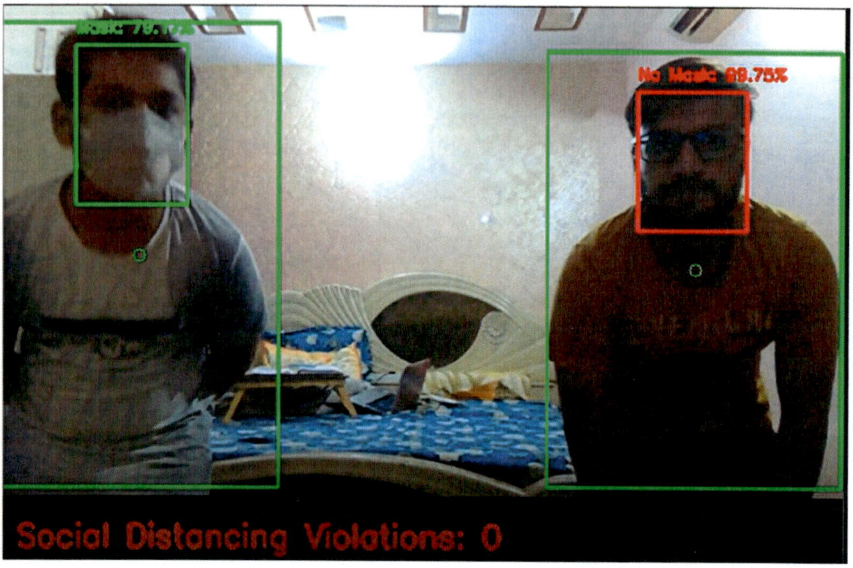

(a)

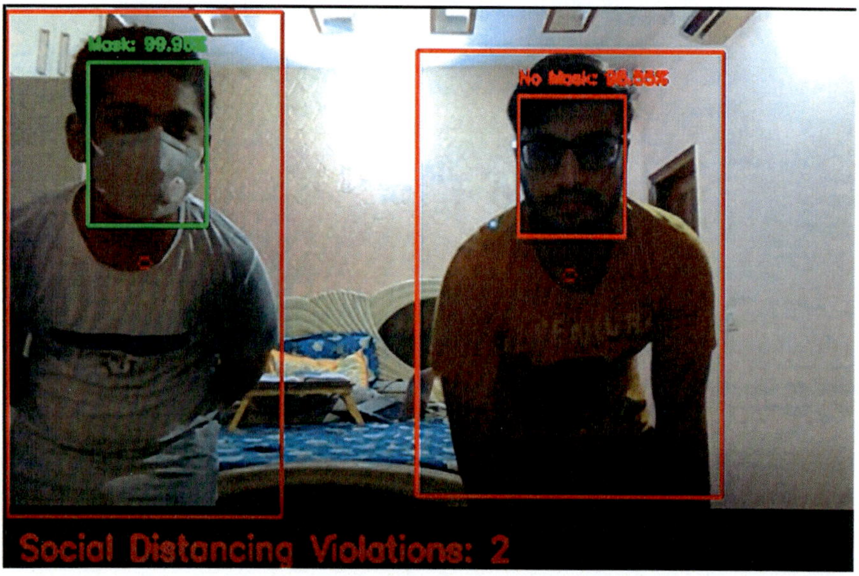

(b)

Fig. 4. (a) When at optimum distance from the other person. (b) When the distance between is less than the threshold

distance and are bounded by two boxes: one for the face mask and other for the body. Similarly, as face mask detection red box represents violation whereas green represents non-violation.

5 Conclusion

In summary, the proposed model can be used in welfare of the society in keeping people safe. Also, the model is very well calibrated to give accurate predictions and help the authorities in identifying the violators. With such an effectiveness, this method is much more accurate and more useful than other models proposed which either detect only face mask or social distancing rather than detecting them altogether.

Acknowledgement. This research paper helped us learning about different concepts of Computer Vision, Deep Learning, Custom Convolutional Neural Networks and Transfer Learning. I would like to give special thanks Mrs. Princy Randhawa for guidance and mentoring me.

References

1. CHAPTER 4 4.: Methods for measuring distance in images 4. (n.d.). www.verypdf.com
2. Fact-check: How much do masks reduce coronavirus ‘contagion probability’? - News - Austin American-Statesman - Austin, TX. (n.d.). 8 September 2020. https://www.statesman.com/news/20200706/fact-check-how-much-do-masks-reduce-coronavirus-lsquocontagion-probabilityrsquo
3. Jia, Y., et al.: Caffe: convolutional architecture for fast feature embedding. In: MM 2014 - Proceedings of the 2014 ACM Conference on Multimedia, pp. 675–678 (2014). https://doi.org/10.1145/2647868.2654889
4. Redmon, J., Divvala, S., Girshick, R., Farhadi, A. (n.d.).: You Only Look Once: Unified, Real-Time Object Detection. http://pjreddie.com/yolo/
5. Sandler, M., Howard, A., Zhu, M., Zhmoginov, A., Chen, L.-C.: MobileNetV2: inverted residuals and linear bottlenecks. In: Proceedings of the IEEE Computer Society Conference on Computer Vision and Pattern Recognition, pp. 4510–4520. http://arxiv.org/abs/1801.04381
6. Viola, P., Jones, M.: Rapid object detection using a boosted cascade of simple features. In: Proceedings of the IEEE Computer Society Conference on Computer Vision and Pattern Recognition, pp. 1. https://doi.org/10.1109/cvpr.2001.990517
7. Zou, Z., Shi, Z., Guo, Y., Ye, J. Object detection in 20 years: a Survey. http://arxiv.org/abs/1905.05055

Evaluation of Voltage Stability Indices

Adebola Soyemi[1(✉)], Sanjay Misra[1,2], Jonathan Oluranti[1], and Ravin Ahuja[2]

[1] Center of ICT/ICE Research, Covenant University, Canaanland, Nigeria
{adebola.soyemi,sanjay.misra,
jonathan.oluranti}@covenantuniversity.edu.ng
[2] Shri Viswakarama Skill University, Gurgaon, India

Abstract. Voltage security/stability appraisal and control are not regarded as new issues. Nevertheless, they rather gained unusual attention to preserve the stability of power transmission networks and evade repeat of major power outages as experienced in some countries(like United States, Canada, Belgium, Sweden, Tokyo, Tennessee). Voltage stability evaluation is indispensable in monitoring power system stability. For ten (10) years, the Nigeria National Gird (NNG) has experienced a total of 29.3 collapses. This work demonstrates a comparison of six voltage stability indices referred to as a line (i.e., Lmn, FVSI, LQP, Lp, NVSI, and NLSI_1); it shows their advantages and disadvantages. The effectiveness of these indices is evaluated via numerical studies in the IEEE 14-bus test system under diverse loading situations. From the study of these indices, a suitable index would be chosen to monitor the Nigerian power system.

Keywords: Voltage stability study · Voltage security · Line stability index

1 Introduction

In power systems, the problems of voltage security are a cause of significant worries to the professional for operation and planning. Voltage stability was described by [1] as the capability of a system to retain adequate voltages at all buses in the system during normal operation and, later, subject to a disturbance. The phenomenon of voltage collapse has been observed in several countries, including Nigeria. According to available data, for over ten (10) years, the Nigeria National Gird (NNG) was seen to have an annual rate of collapse of 29.3 [4].

Voltage stability studies can be examined using either dynamic or static approaches. The dynamic approach employs nonlinear differential and algebraic equations in the power system model. These equations are solved through transient stability simulations, and the system response over a certain period of time is observed. The dynamic approach is suitable for large disturbances and therefore important for system control, while the Static approach employs only algebraic equations because it assumes the system is in steady-state. This approach is suitable for small disturbances in the system because of this; it uses the conventional power-flow model.

Even though stability studies generally require dynamical models of the power system but in most literature, the study is approached using static techniques [2–4]. The

L. Garg et al. (Eds.): ISMS 2020, LNNS 303, pp. 74–82, 2022.
https://doi.org/10.1007/978-3-030-86223-7_8

static approach makes it easier to model and analyze the system because it is assumed to be in a steady-state.

Voltage stability indices show how faraway load buses are from their voltage stability limit, and therefore the most critical bus can be identified through maximum load ability [2]. A number of techniques are created to analyze it, e.g., P-V and Q-V curves, sensitivity analysis [5], Jacobian determinant [6], modal analysis [7], and line index [3, 5–8]. P-V, Q-V curves are not used often because they are computationally expensive and do not provide adequate, useful statistics towards stability problems [9, 10] whereas in a power system, voltage stability indices offer dependable information about closeness to voltage collapse. These indices typically have values varying between 0 and 1 [11] where 0 represents no lad and one as the collapse of voltage.

In a power system, voltage stability indices are obtained with respect to either a bus or line. In this paper, six voltage stability indices are obtained with respect to a line, the performance capability of each index in identifying the point of voltage instability or collapse, load-ability limits, and critical line in the system during different loading cases is reviewed. The load is gradually increased from the base until the indices reach a value close to 1; which indicates voltage collapse. The outcomes of simulating on IEEE 14-bus will discourse. All simulations would be done with a program written in a MATLAB environment.

There are researches going on how to save energy, sustainability, and another related area which we are not considering in detail [23–27] because they are not much related to our work.

Voltage stability index formulation is discussed in Sect. 2. Comparison of stability indices has discoursed in Sect. 3. Test results, identification of weak bus, and the most critical line is explained in Sect. 4—Finally, conclusions in Sect. 5.

2 Voltage Stability Indices and Its Formulation

The voltage security/stability indices can be obtained with respect to either a bus or line. A stability study can be carried out on systems by assessing the voltage stability index. The six voltage stability indices obtained wrt a line presented in this paper are; LQP, Lmn, FVSI, NVSI Lp, and NLSI_1. LQP, Lmn, NVSI, Lp, and FVSI are developed as in Fig. 1, while NLSI_1 is formulated from the combination of Lmn and FVSI indices.

The stability criterion of these indices is obtained by setting the voltage quadratic equation discriminator. Figure 1 shows a single line diagram of an interconnected system.

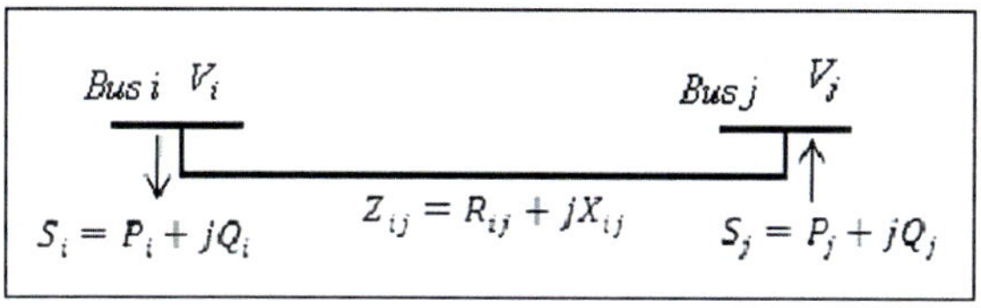

Fig. 1. Two bus model

1. Line Stability Index Lmn: In [12], the authors formulated the line stability index using the system in Fig. 1. The index is centered on power transmission theory in a single line using pie (π) model representation. To maintain stability, the index Lmn should be less than 1. Lmn [12] is defined as:

$$\text{Lmn} = \frac{4XQ_2}{[|V_1|\sin(\theta - \delta)]^2} \leq 1 \quad (1)$$

Where θ is line impedance angle, X is line reactance, Q_2 is reactive power flow at the receiving end, V_1 is the voltage at the sending end, and δ is angle difference between the voltage at the sending and receiving ends.

2. Line Stability Index FVSI: FVSI put forward by the authors in [13] is centered on the flow power through a single line. To maintain stability, FVSI must be less than 1. The FVSI index [13] is expressed as:

$$\text{FVSI} = \frac{4|Z|^2|Q_2|}{|V_1|^2 X} \leq 1 \quad (2)$$

Where Z is line impedance, X is line reactance, Q_2 is reactive power flow at the receiving end, and V_1 is the voltage at the sending end.

3. Line Stability Index LQP: It is obtained by authors in [14] and created using the same concept as in [12, 13]. For the system to remain stable, the value of this index should be less than 1. The LQP [14] is expressed as:

$$\text{LQP} = \left(\frac{4X}{V_1^2}\right)\left(\frac{X}{V_1^2}P_1 + Q_2\right) \leq 1 \quad (3)$$

Where X is line reactance, Q_2 is reactive power flow at the receiving bus, V_1 is the voltage of the sending bus, and P_1 is real power flow at the sending bus.

4. On-Line Voltage Stability Index (L_p): This index [15] shows the interrelationship of active power and bus voltage. To maintain stability, this index should be less than 1. The index is [16] expressed as:

$$\text{Lp} = \frac{4RP_2}{(V_1\cos(\theta - \delta))^2} \leq 1 \quad (4)$$

Where R is line resistance, V_1 is the voltage of the sending bus, and P_2 is the real power flow at the receiving bus.

5. New Voltage Stability Index NVSI: Authors in [3] obtained NVSI by assuming the system in Fig. 1 has a transmission line resistance of zero. To maintain stability, NVSI should be less than 1. NVSI [3] is mathematically expressed as:

$$\text{NVSI} = \frac{2X\sqrt{P_2^2 + Q_2^2}}{2Q_2X - |V_1|^2} \quad (5)$$

Where X is line reactance, P_2 is real power flow at the receiving end, Q_2 is reactive power flow at the receiving end, and V_1 is the voltage at the sending end.

6. Newline stability index NLSI_1: NLSI-1 proposed in [4] is derived from the combination of Eqs. (1) and (2). Closeness to voltage instability is computed according to a switching function, σ given by Eq. (6). To maintain stability, NLSI_1 must be less than 1. NLSI-1 [4] is expressed as:

$$\mathrm{NLSI\_1} = \frac{4Q_2}{|V_1|^2}\left[\frac{(|Z|^2)}{X}\sigma - \frac{X}{Sin^2(\theta - \delta)}(\sigma - 1)\right] \leq 1 \quad \sigma = \begin{cases} 1\delta < \delta_c \\ 0\delta \geq \delta_c \end{cases} \tag{6}$$

Where Z is line impedance, X is line reactance, Q_2 is reactive power flow at the receiving end, and V_1 is voltage at the sending end, θ is line impedance angle, δ is angle difference between the sensing end voltage and the receiving end voltage and σ is the switching function whose value is dependent on whether or not angle difference δ is small.

3 Comparison of Stability Indices

Lmn makes no other assumption other than shunt resistance been zero. The major advantage of Lmn is its insensitivity to the resistance-reactance ratio of the transmission line as shown by the author in [17] and its drawback is that does not account for the flow of real power of the line in its forecast of voltage stability and hence might be inaccurate under certain operating conditions [8]. On the other hand, when compared with indices such as FVSI and LQP, it is more responsive to real power changes because of its indirect connection to real power through voltage angle difference δ.

FVSI assumes that the angle difference between the voltages between sending and receiving ends is approximatively zero. This assumption is one of the drawbacks of this index, as large-angle differences could be a pointer to the occurrence of collapse [4, 18]. Another drawback is its sensitivity to the resistance-reactance ratio of the transmission line [17]. Its advantage is that it is considerably fast.

LQP assumes that R/X $< < 1$, it also disregards the resistance of the transmission line, which could cause it to be inaccurate under certain operating conditions. The advantage of this index is it's insensitive to the resistance-reactance ratio of the transmission line [17].

Lp assumes the effect of reactive power is negligible. This index has the following drawbacks: 1.) it fails if the transmission line resistance is very close to zero, 2.) it is sensitive to δ, which can lead healthy lines to be identified as a critical line because $\cos(\theta - \delta)$ is faster than $\sin(\theta - \delta)$ around 90º [4, 19] and 3.) Its accuracy is greatly affected whenever the load power factor is low, which is less than 0.80, and for this reason, it can only be used in distribution systems that are mainly resistive [15]. It is easy to implement.

NVSI assumes the transmission line resistance is approximately zero. Its drawback is that adversely affected by the resistance-reactance ratio of the transmission [17]. It has the advantage of being able to monitor a power system in real-time [3].

NLSI_1makes no other assumptions, as the assumption made by one of its parent index (FVSI) has been corrected through the introduction of the switching function σ -which switches to Lmn once the angle difference is large. The drawback of this index is its sensitivity to the resistance-reactance ratio of the transmission line when the angle difference is small. Its major advantage is its increased accuracy and speed.

Table 1, at a glance, shows each index's attributes and highlights their advantages and drawbacks. From the mathematical expression of each index, a relationship can be seen between the indices; in that, some index can be derived from others' simplification. For instance, FVSI can be obtained from Lmn by setting δ to zero.

4 Test Results and Identification of Weak Bus and Critical Lines

4.1 Testing System and Tool for Analysis

The investigation into the performance of the stability indices is performed on IEEE 14-Bus. The system configuration of the test system is shown in Fig. 2. The system is comprised of five generator buses (PV), nine load buses (PQ), and twenty interconnected lines. Figure 2 shows the single-line diagram of the system. The aim of simulations was to compare the operation of the different indices used to detect proximity to voltage instability or collapse under different loading conditions.

To carry the analysis, a bus is chosen at random and each index's performance is assessed under two case studies; reactive power and active power load changes. The load at the selected bus is increased till it is close to its voltage stability limit. With respect to the load change the line with highest index value is termed the most critical line and the bus, a weak bus. This is repeated for each index.

a. **Reactive Power Load Changes**

This loading scenario determines the maximum load-ability and critical line of all the load buses by varying reactive load on each bus until the index value approaches 1. The reactive power of the load buses were changed one at a time to examine the maximum reactive power on each bus.

A weak bus is defined as the bus that has the lowest load-ability limit. This bus therefore requires compensation devices to avert and mitigate against voltage collapse or instability. The result of this analysis is shown in Table 4.

From Table 2 and 3, bus 14 is identified as a weak bus because it has the lowest load-ability limit and its most critical line is identified as line 13–14.

b. **Real Power Load Changes**

Consequence of real power loading on the indices is examined by increasing real power at each bus until the verge of voltage collapse. From Table 4, Lq and NVSI reach unity faster than the other indices while other indices are lagging far behind them.

Table 1. Comparisons of voltage stability indices

Index	Expression	Assumption	Critical Value	Drawbacks	Advantages
Lmn	$\frac{4XQ_r}{[\lvert V_S\rvert \sin(\theta-\delta)]^2}$	None	1.00	It may become inaccurate under certain operating conditions since does not account for the real power flow of the line [8]	It can be used for lines having high Resistance-Reactance ratios
FVSI	$\frac{4\lvert Z\rvert^2 \lvert Q_2\rvert}{\lvert V_1\rvert^2 X}$	$\delta \approx 0$	1.00	The angle difference δ is assumed to be zero [4, 18]. It cannot be used for transmission lines with a high Resistance-Reactance ratio	It is considerably fast
LQP	$\left(\frac{4X}{V_s^2}\right)\left(\frac{X}{V_s^2}P_s + Q_r\right)$	R/X < < 1	1.00	It could give inaccurate results under certain operating conditions since it ignores the resistance of transmission lines	It can be used for lines having high Resistance-Reactance ratios
Lp	$\frac{4RP_r}{(V_S \cos(\theta-\delta))^2}$	The effect of reactive power is neglected	1.00	The index fails if the resistance R is close to zero Highly sensitive to δ Accuracy is affected by the power factor	It is easy to implement
NVSI	$\frac{2X\sqrt{P_2^2+Q_2^2}}{2Q_2X-\lvert V_1\rvert^2}$	$R \approx 0$	1.00	It is adversely affected by the R/X ratio of the transmission line	It can be used for real-time monitoring
NLSI_1	$\frac{4Q_r}{\lvert V_s\rvert^2}\left[\frac{(\lvert Z\rvert^2)}{X}\sigma - \frac{X}{Sin^2(\theta-\delta)}(\sigma-1)\right]$	None	1.00	It is affected by the R/X ratio of the transmission line when angle difference δ is small	It also has better accuracy and speed

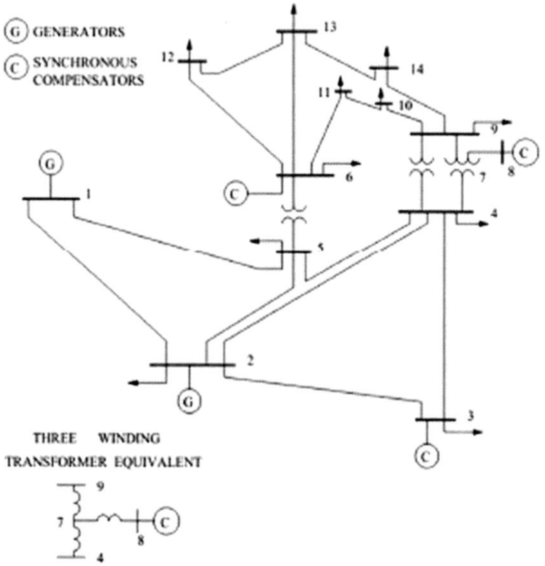

Fig. 2. Single-line diagram of the IEEE 14-Bus System

Table 2. Line VSI under heavy reactive loading

Bus No.	Loading (p.u.)	Critical Line	Lmn	LQP	FVSI	Lp	NVSI	NLSI_1
4	3.610	3–4	0.947	0.931	1.045	1.030	0.813	1.045
7	1.654	7–8	0.994	0.978	0.994	0.962	0.800	0.994
10	1.217	10–11	0.998	0.980	0.955	0.969	0.880	0.998
14	0.755	13–14	0.923	0.905	0.974	0.998	0.900	0.923

Table 3. Line VSI under heavy real loading

Bus No.	Critical Line	Lmn	LQP	FVSI	Lp	NVSI	NLSI_1
4	3–4	0.534	0.514	0.513	0.833	0.800	0.513
7	7–8	0.670	0.670	0.621	0.898	0.813	0.621
10	10–11	0.899	0.701	0.710	0.935	0.911	0.899
14	13–14	0.834	0.835	0.785	1.010	0.985	0.834

5 Conclusion

This paper presented six voltage stability indices referred to as a line. It made comparisons of each index in terms of their advantages, drawbacks, and performance under different loading conditions. From the results obtained through simulation, it is observed that Lmn, FVSI, LQP, and NLSI_1 are responsive to changes in reactive power while Lq and NVSI are responsive to changes in real power.

All six indices could be used to monitor voltage stability for the Nigerian grid, but some of them were found not suitable due to their limitations (drawback). For instance, L_p cannot be used in systems where the power factor is low. NVSI is also not suitable also because its accuracy is also affected by the power factor of the system.

NLSI_1 is better suited to monitor voltage stability in the Nigerian grid because it combines the advantages of two indices through a switching function σ, which is used to switch quickly between the two indices.

Acknowledgment. The authors appreciate the sponsorship from Covenant University through its Centre for Research, Innovation and Discovery, Covenant University, Ota Nigeria.

References

1. Kundur, P.: Power System Stability and Control, McGrew-Hill (1994)
2. Jalboub, M., Rajamani, H., Liang, T., Abd-Alhameed, R., Ihbal, A.: Investigation of voltage stability indices to identify weakest bus (TBC). In: Mobile Multimedia Communications. MobiMedia 2010, Springer, Berlin, Heidelberg, pp. 682-687 (2012)
3. Kanimozhi, R., Selvi, K.: A novel line stability index for voltage stability analysis and contingency ranking in power system using fuzzy based load flow. J. Electr. Eng. Technol. **8**(4), 694–703 (2013)
4. Samuel, I.: A New Voltage Stability Index For Predicting Voltage Collapse In Electrical Power System. Covenant University Repository, Ota (2017)
5. Yorino, N., El-Araby, E., Sasaki, H., Harada, S.: A new formulation for FACTS allocation for security enhancement against voltage collapse. IEEE Trans. Power Syst. **18**(1), 3–10 (2003)
6. Kazemi, A., Shayanfar, H., Rabiee, A., Aghaie, J.: Power system security improvement using the unified power flow controller. In: IEEE Power India Conference, India (2006)
7. Chang, Y.-C.: Multi-objective optimal SVC installation for power. IEEE Trans. Power Syst. **27**(2), 984–992 (2012)
8. Ratra, S., Tiwari, R., Niazi, K.: Voltage stability assessment in power systems using line voltage stability index. Comput. Electr. Eng (2018). https://doi.org/10.1016/j.compeleceng.2017.12.046
9. Balamourougan, V., Sidhu, T., Sachdev, M.: Technique for online prediction of voltage collapse. IEEE Proc. Gener. Transm. Distrib. **151**(4), 453–460, 11 July 2004
10. Goh, H., Chua, Q., Lee, S., Kok, B., Goh, K., Teo, K.: Evaluation for voltage stability indices in power system using artificial neural network. Procedia Eng. **118**, 1127–1136 (2015)
11. Reis, C., Barbosa, F.M.: A comparison of voltage stability indices. In: IEEE MELECON, Benalmadena (Malaga), Spain (2006)
12. Moghavemmi, M., Omar, F.: Technique for contingency monitoring and voltage collapse prediction. IEE Proc. Gener. Transm. Distrib. **145**(6), 634–640 (1998)
13. Musirin, I., AbdulRahman, T.: Novel fast voltage stability index (FVSI) for voltage stability analysis in power transmission system. In: Student Conference on Research and Development, SCOReD, Shah Alam, Malaysia (2002)
14. Mohamed, A., Jasmon, G., Yusof, S.: A static voltage collapse indicator using Line Stability Factors. J. Ind. Technol.**7**(1), 73-85 (1998)
15. Moghavvemi, M., Faruque, M.: Technique for assessment of voltage stability in ill-conditioned radial distribution network. IEEE Power Eng. Rev. **21**(1), 58–60 (2001)

16. Suganyadevia, M.V., Babulalb, C.K.: Estimating of loadability margin of a power system by comparing Voltage Stability Indices. In: International Conference on Control, Automation, Communication and Energy Conservation, Perundurai, Tamilnadu (2009)
17. Yari, K.H.S.: Assessment of line stability indices in detection of voltage stability status. In: IEEE International Conference on Environment and Electrical Engineering and IEEE Industrial and Commercial Power Systems Europe (EEEIC/I&CPS Europe), Milan (2017)
18. Dobson, I., Parashar, M., Carter, C.: Combining phasor measurements to monitor cutset angles. In: 43rd Hawaii International Conference on System Sciences (HICSS) (2010)
19. Haruna, M.: An overview on voltage stability indices as indicators of voltage stability for networks with distributed generations penetration. Int. J. Sci. Technol. Soc. **3**(4), 214–219 (2015)
20. Chow et al.: On the evalution of voltage collapse criteria. IEEE Trans., PWRS-5, 612–620 (1990)
21. U. Nations: Sustainable development goal 7. http://www.un.org/sustainabledevelopment/energy/
22. Kamel, S., Kodsi, M., Canizares, C.: Modeling and simulation of IEEE 14 bus system with facts controllers. University of Waterloo, Canada, Tech. Rep (2003)
23. Samuel, I., Katende, J., Daramola, S., Awelewa, A.: Review of system collapse incidences on the 330-kV Nigerian national grid. Int. J. Eng. Sci. Invention **3**, 55–59 (2014)
24. Adesanya, A., Misra, S., Maskeliunas, R., Damasevicius, R.: Prospects of ocean-based renewable energy for West Africa's sustainable energy future. Smart and Sustain. Built Environ. (2020)
25. Ayodele, E., Misra, S., Damasevicius, R., Maskeliunas, R.: Hybrid microgrid for microfinance institutions in rural areas–a field demonstration in West Africa. Sustain. Energy Technol. Assess. **35**, 89–97 (2019)
26. Ogunleye, O., Alabi, A., Misra, S., Adewumi, A., Ahuja, R., Damasevicius, R.: Comparative study of the electrical energy consumption and cost for a residential building on fully AC loads Vis-a-Vis one on fully DC loads. In: Advances in Data Sciences, Security and Applications, pp. 395–405, Springer, Singapore (2020)
27. Aderemi, O., Misra, S., Ahuja, R.: Energy consumption forecast using demographic data approach with Canaanland as case study. In: International Conference on Next Generation Computing Technologies, pp. 641–652, Springer, Singapore (2017)
28. Adeyemi-Kayode, T.M., Misra, S., Damaševičius, R.: Impact analysis of renewable energy based generation in West Africa – a case study of Nigeria, Problemy Ekorozwoju – Problems of Sustain. Dev. **16**(1), 67–78 (2021)

Optimizing Fused Deposition Modelling Process Parameters Using Metaheuristic Machine Learning Algorithms

Jatin Deep Kharbanda, Yakshrat Nanda, Gireesh Dangayach, and D. A. P. Prabhakar(✉)

Department of Mechatronics, Manipal Institute of Technology, Manipal Academy of Higher Education (MAHE), Manipal, India
pavan.prabhakar@manipal.edu

Abstract. In recent times, the Additive Manufacturing technique FDM or Fused Deposition Manufacturing, has grown to become a widely used method to produce complicated parts in a rapid layer-by-layer manufacturing process. This results in a lower production cycle time when compared to other manufacturing processes. FDM construction is affected by several process parameters that impact build time, mechanical properties, part quality, and dimensional accuracy. Some of these parameters are air gap, build orientation, infill percentage, raster angle, layer thickness, etc. whose importance varies under different circumstances of manufacture. The process parameters need to be selected carefully for each part. For a specific output requirement, some of the process parameters are more significant than the rest. Through a careful literature study, the four most important parameters have been determined and translated into an objective equation. One widely used method of optimization for other additive manufacturing techniques is to involve Metaheuristic optimization algorithms to obtain the closest approximation to an ideal value for the significant parameters. Metaheuristic algorithms will soon be an integral part of modern optimization. Optimization is to be done through MATLAB to obtain the ideal values of necessary parameters for the most optimal output.

Keywords: Fused deposition modelling (FDM) · Metaheuristic · Optimization · Process parameters

1 Introduction

Additive Manufacturing (AM) describes a technology that employs a method to build objects by adding consecutive layers of material in a 3D environment. AM commonly involves the use of 3D modelling Computer-Aided Design (CAD) software, machinery, and construction material. Once a CAD sketch is produced, the AM equipment interprets the file as layers to construct. It then adds consecutive layers of a sheet, powder, or liquid material or otherwise sequentially to fabricate a 3D object. Several methods work on a varied basis. However, these processes are similar in terms of the thermal, mechanical, and chemical ways they construct workpieces.

L. Garg et al. (Eds.): ISMS 2020, LNNS 303, pp. 83–92, 2022.
https://doi.org/10.1007/978-3-030-86223-7_9

The most used processes are Stereolithography, Fused Deposition Modelling, Ballistic Particle Manufacturing, and Selective Laser Sintering. AM processes are separated into solid, liquid, and powder-based types. As their names suggest, they work with solid, liquid, and powder feedstock. Some of the benefits of AM processes are:

i. No tooling design is required
ii. Singular machine for multiple parts
iii. Lower wasted materials and the final cost

Fused deposition modelling is the primary focus of this study. It is an additive process that utilizes polymers for manufacturing. This setup is equipped with a computer-controlled nozzle head that deposits semisolid materials on a surface to form a layer. This process is widely used for hard polymers as they exist in a semisolid form before deposition.

The deposition materials exit the nozzle head as a filament with geometric and dimensional variance depending on the nozzle head parameters. The fabrication takes place layer by layer. After the deposition of each layer, a milling head cuts the surface to adjust the thickness and finish of the layers. To facilitate material deposition, FDM occurs at temperatures near the melting point of the deposition materials. Additionally, a second nozzle may be added to the setup to deposit secondary materials to create composite parts. This can modify the properties of the fabricated components. Sensor based data may also be used to monitor progress to allow for machine learning optimization [1].

A benefit of using this process is to obtain the finished part at the end without the need for machining and processing [2]. Certain parameters involved in the fabrication process are more involved in the quality of the final output, for example Bai Huang et al. [3] focused upon four processing parameters. The parameters were printing speed, layer thickness, raster angle, and building orientation. They were analysed in terms of their mechanical properties, the microstructure of acrylonitrile butadiene-10 styrene (ABS), and surface quality.

Kira Barton et al. [4] focused on designing a printing clarity that eradicates the elevated rate of collapses of Additive manufacturing processes, multi-material manufacturing, and embedded actuation and sensing, requiring the unification of the contrasting materials. Yicha Zhang et al. [5] focused on the build time estimation for the parts of different designs, optimization, and price citation in Additive Manufacturing (AM) also focusing on an integrated adaptive modelling method that was procured from Grey Theory. The average estimation accuracy of the model is within 10%, which, when compared to other parametric and analogical models, is remarkable.

2 Optimization

A key factor in this process is the requirement of a CAD model to obtain the slices necessary to fabricate the parts. The control units of the fabrication machines operate at user-defined settings to produce parts of a specific quality. There is still room to optimize this process to ensure that maximum use is obtained from the designed parts. L. Villalpandoa et al. [6] built components having parametric internal structures with fused deposition modelling using the optimization approach. The research tells us about the varied methods for outlining the characteristics of the different types of substances used in the printing procedure. One of the methods is through conventional mechanics of substances, using force equilibrium and displacement continuity to demonstrate the standard macroscopic stress and strain states. There are some primary theories in modelling unidirectional composites.

Machine learning is a method to automate improvement within manufacturing that has been significantly researched upon [7]. Use of artificial intelligence has also been explored to determine the most significant factors [8], Hyunwoong Ko, Paul Witherell, Ndeye Y. Ndiaye, and Yan Lu et al. [9] elaborated on the algorithm to create predictive models that correlate part structure and process parameters. Moreover, it also focuses on using a few ML-based algorithms to improve the design, process control, in-situ monitoring, and ex-situ measurement. With the help of the algorithm, AM programs can improve autonomously, while displaying continuously generated effective AM data, in an autonomous and automated approach.

Sandeep Deswal et al. [10] experimented on the modelling and parametric enhancement of the FDM 3D-printing process using composite techniques to amplify the dimensional precision. The study mainly focused parameters layer thickness, infill density, build orientation, and the number of contours. For the training and optimization purposes, hybrid statistical tools and artificial neural network-genetic algorithms (ANN-GA), and genetic algorithms in MATLAB were used.

One widely utilized method of optimization for additive manufacturing techniques is using Metaheuristic optimization algorithms to obtain the closest approximation to an ideal value for the most significant parameters. Metaheuristic algorithms will soon become an integral part of modern optimization strategies. Arup Dey et al. [11] attempted to optimize the FDM process parameters through the utilization of the PSO Algorithm. This was done to obtain a lower build time and higher compressive strength, four parameters: infill density, build orientation, layer thickness, and extrusion temperature are optimized. It is determined that the extrusion temperature is the most insignificant parameter for both the required outcomes.

A metaheuristic is defined as a high-level heuristic solution designed to find, generate, or select a partial search algorithm or heuristic that is the most likely to generate an optimized solution to a problem with imperfect or incomplete information. Modern optimization algorithms are often inspired by natural phenomena. The inspiration is diverse and consequently, the algorithms are as well.

However, all algorithms draw from some specific characteristics to determine the key update formula. Darwinian evolutionary characteristics of biological systems inspired the genetic algorithms which apply the concepts of mutation and survival of the fittest. The solutions are represented as chromosomes to be modified. Particle swarm optimization imitates the swarming behaviour of group animals such as birds or fish, as they demonstrate characteristics of swarm intelligence. In this study, Firefly optimization is utilized.

Standard Fused Deposition Manufacturing (FDM) techniques are not very efficient towards the material used. Due to these drawbacks, the applications of the FDM process are limited, for example, high built time, poor surface quality, or poor mechanical properties.

Ashay Kohad, Rajendra Dalu et al. [12] identified the distinct process parameters affecting the quality of the manufactured parts with FDM. This study found how different performance parameters are affected based on earlier research. This study also compares different optimization techniques and identifies the most suitable technique for the enhancement of process parameters in FDM 3D printing. The same techniques on other types of additive manufacturing have been explored as well [13].

The objective of this paper is to determine the optimal values of selected process parameters for a Fused Deposition Modelling (FDM) process using Machine Learning Algorithms and highlight the relationship between layer thickness and build orientation. The paper is organized as follows: an inquiry alighting from starting conditions which defines the existing problem, a methodology section that discusses the procedure to identify the process and critically evaluate the validity, the experimental design used in the study, followed by the results and conclusion.

3 Methodology

Identify research done in a similar vein to the problem faced and extract necessary information on gaps in research and potential solutions. Identifying the most significant parameters involved in the process through literature reviews then determine the parameters to be optimized. This is followed by developing an objective function that governs the outcome of the printed object using the most significant parameters as the major factors. This functional equation will be used to develop the firefly algorithm which will be applied to determine the optimized values for each parameter then comes time to test optimized parameters on a printed standard piece and perform a character study of the optimized and unoptimized workpiece to determine quality of improvement. Figure 1 shows the methodology followed in a flow chart.

4 Experimental Design

The present experimental design aims to achieve the maximum possible amount of information from the smallest number of experiments performed. For this study, the effects of four variables were studied, and a total of 30 permutations of process parameters were generated.

The work has been done with a quadratic model for compressive strength that represented the relationship of compressive strength with the different process parameters, namely layer thickness (x_1) (mm), build orientation (x_2) (Degree), infill density (x_3) (%), and extrusion temperature (x_4) ($^{\circ}$C).

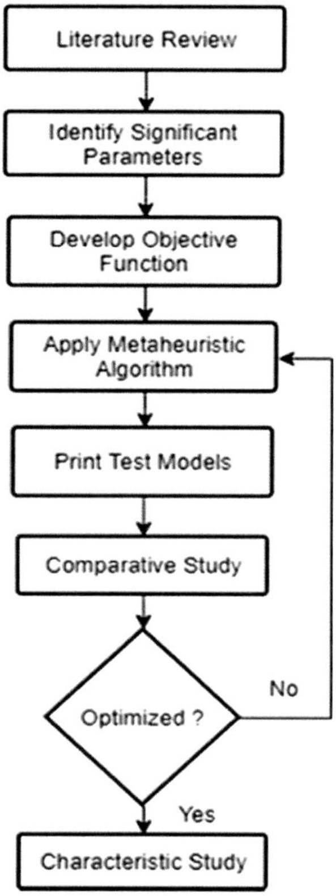

Fig. 1. Flow chart of methodology

The equation for compressive strength is given below:

$$
\begin{aligned}
\textit{Compressive Strength} &= 16.679 + 1.217x_1 + 1.483x_2 + 3.216x_3 + 7.430x_2 \\
&- 3.418x_1x_2 + 2.910x_1x_3
\end{aligned}
\tag{1}
$$

We are setting the lowest values of x2 and x3 as −1, and solving the compressive strength equation for x1 using the Firefly algorithm. By using the Firefly algorithm, we are finding the global minimum for x1 and comparing it with values obtained in the table. The experimental result of x1 should tend to −1.

5 Results and Discussion

In Fig. 2 below, the blue points indicate the initial iteration of Fireflies spawned at random locations. Each subsequent color grouping displays the new location and clustering of the fireflies through each iteration. It is observed that the fireflies group together towards the ideal global minima, in this case, x1 = −1.

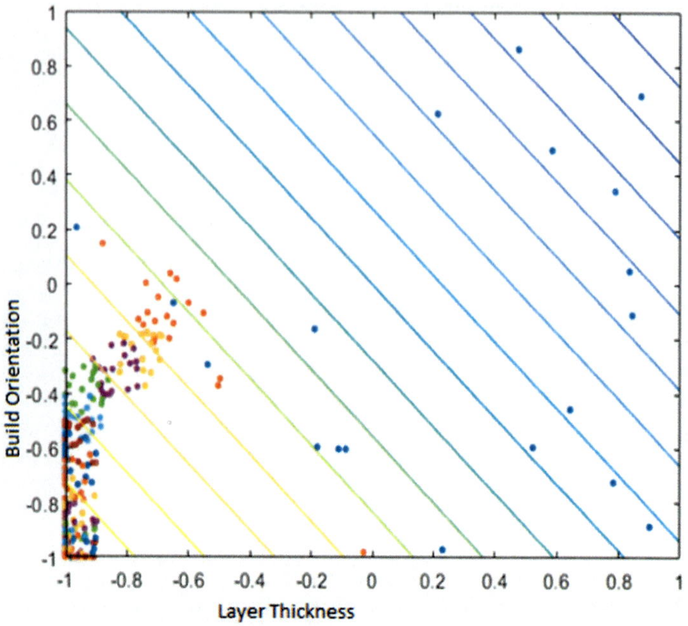

Fig. 2. Plot representing the relation between layer thickness and build orientation. The x-axis is x_1 or layer thickness and the y-axis is x_2 or build orientation.

Figure 3 determines the relationship between the layer thickness and build orientation using the algorithm to determine the ideal compressive strength over a variety of settings applied to x1, x2, and x3 along with their range of variance (Table 1).

The final optimal value as per simulation is when compressive strength is equal to 33.387 in the scenario where the Layer thickness and Infill density are maximized to the degree possible on each 3D printing device, while build orientation is minimized to zero degrees.

Table 1. Effect of factors on compressive strength.

Serial Number	Layer Thickness, X_1 (mm)	Build Orientation, X_2 (Degree)	Infill Density, X_3 (%)	Compressive Strength
1	−1	−1	−1	16.6850
2	−1	−1	1	17.2970
3	−1	1	−1	20.6670
4	−1	1	1	27.0990
5	1	−1	−1	21.1350
6	*1*	*−1*	*1*	*33.3870*

(*continued*)

Table 1. (*continued*)

Serial Number	Layer Thickness, X_1 (mm)	Build Orientation, X_2 (Degree)	Infill Density, X_3 (%)	Compressive Strength
7	1	1	−1	17.2650
8	1	1	1	29.5170
9	−1	0	0	15.4620
10	0	−1	0	22.6260
11	0	0	−1	13.4630
12	1	0	0	17.8960
13	0	1	0	25.5920
14	0	0	1	19.8950

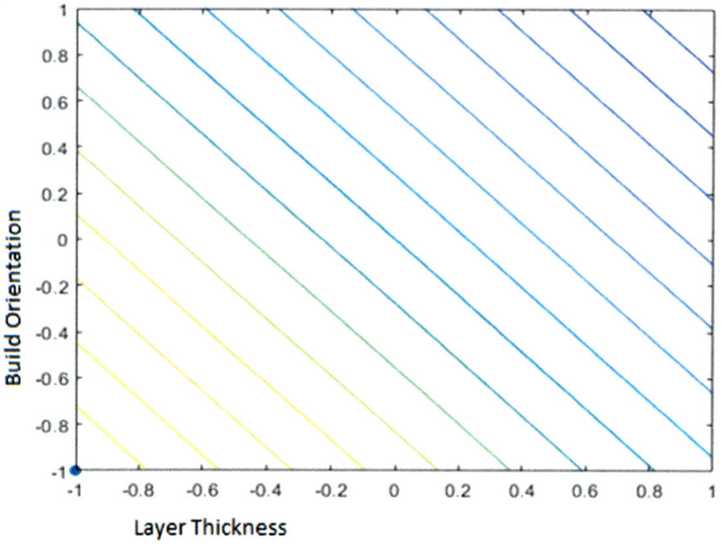

Fig. 3. The output using the algorithm to determine the ideal compressive strength over a variety of settings applied to x_1, x_2, and x_3 along with their range of variance. The x-axis is x_1 or layer thickness and the y-axis is x_2 or build orientation.

6 Conclusion

A literature review was done for understanding the effects of each process parameters on the FDM process. The firefly algorithm was used to run the equation for compressive strength. A relationship between layer thickness and build orientation was plotted using a MATLAB code.

Annexure

Firefly Code:

```
function [best]=firefly_simple(instr)
if nargin<1, instr=[50 100];
End
n=instr(1); MaxGeneration=instr(2);
help firefly_simple.m
rand('state',0);
g=10;
X=-1;



% Function Definition%

function [best]=firefly_simple(instr)
if nargin<1,   instr=[20 100];
end
n=instr(1);  MaxGeneration=instr(2);
help firefly_simple.m
rand('state',0);
g=100;
X=-1;
% Function Definition%

str1='-(16.679+(1.217)+3.21*X)';
str2='-(-1.483+7.430+3.418*X-2.910*X+g)';

funstr=strcat(str1,str2);
f=vectorize(inline(funstr));
range=[-1 1 -1 1];
alpha=0.2;
gamma=1.0;
delta=0.97;
% Grid values used for display%
Ngrid=100;
dx=(range(2)-range(1))/Ngrid;
dy=(range(4)-range(3))/Ngrid;
[X,g]=meshgrid(range(1):dx:range(2),...
               range(3):dy:range(4));
z=f(X,g);
% Display the shape of the objective function%
figure(1);    surfc(X,g,z);
% generating the initial locations of n fireflies%
[xn,yn,Lightn]=init_ffa(n,range);
```

```
figure(2);
for i=1:MaxGeneration,
contour(X,g,z,15); hold on;
zn=f(xn,yn);
% Ranking the fireflies by their light intensity%
[Lightn,Index]=sort(zn);
xn=xn(Index); yn=yn(Index);
xo=xn;   yo=yn;    Lighto=Lightn;
plot(xn,yn,'.','markersize',20,'markerfacecolor','g');
% Move all fireflies to the better locations%

[xn,yn]=ffa_move(xn,yn,Lightn,xo,yo,Lighto,alpha,gamma,ra
nge);
drawnow;
    hold off;
end
best(:,1)=xo'; best(:,2)=yo'; best(:,3)=Lighto';
% The initial locations of n fireflies%
function [xn,yn,Lightn]=init_ffa(n,range)
xrange=range(2)-range(1);
yrange=range(4)-range(3);
xn=rand(1,n)*xrange+range(1);
yn=rand(1,n)*yrange+range(3);
Lightn=zeros(size(yn));
% Move all fireflies toward brighter ones%
function [xn,yn]=ffa_move(xn,yn,Lightn,xo,yo,...
    Lighto,alpha,gamma,range)
ni=size(yn,2); nj=size(yo,2);
for i=1:ni,

    for j=1:nj,
r=sqrt((xn(i)-xo(j))^2+(yn(i)-yo(j))^2);
if Lightn(i)<Lighto(j),
beta0=1;     beta=beta0*exp(-gamma*r.^2);
xn(i)=xn(i).*(1-beta)+xo(j).*beta+alpha.*(rand-0.5);
yn(i)=yn(i).*(1-beta)+yo(j).*beta+alpha.*(rand-0.5);
end
    end
end
[xn,yn]=findrange(xn,yn,range);
 % Reduce randomness as iterations proceed%
function alpha=newalpha(alpha,delta)
alpha=alpha*delta;
```

```
% Make sure the fireflies are within the range%
function [xn,yn]=findrange(xn,yn,range)
for i=1:length(yn),
   if xn(i)<=range(1), xn(i)=range(1); end
   if xn(i)>=range(2), xn(i)=range(2); end
   if yn(i)<=range(3), yn(i)=range(3); end
   if yn(i)>=range(4), yn(i)=range(4); end
end
```

References

1. Lia, Z., Zhanga, Z.: Prediction of surface roughness in extrusion-based additive manufacturing with machine learning. Robot. Comput.-Integr. Manuf. **57**, 488–495 (2019)
2. Griffiths, C.A., Howarth, J.: Effect of build parameters on processing efficiency and material performance in fused deposition modeling. Procedia CIRP **49**, 28–32 (2016)
3. Huang, B., Meng, S.: Study of processing parameters in fused deposition modeling based on mechanical properties of acrylonitrile-butadiene-styrene filament. Polym. Eng. Sci. **59**(1), 120–128 (2019)
4. Barton, K.: Mechatronics advances for the next generation of AM process control. Int. Fed. Autom. Control **64** (2019). Article ID: 102281
5. Zhang, Y., Bernard, A.: Fast adaptive modeling method for build time estimation in additive manufacturing. CIRP J. Manuf. Sci. Technol. 49–60 (2015)
6. Villalpandoa, C., Eiliata, H.: An optimization approach for components built by fused deposition modeling with parametric internal structures. Procedia CIRP **17**, 800–805 (2014)
7. Razvi, S.S., Feng, S.: A review of machine learning applications in additive manufacturing. In: International Design Engineering Technical Conferences and Computers and Information in Engineering (2019)
8. Douard, A.: An example of machine learning applied in additive manufacturing. In: International Conference on Industrial Engineering and Engineering Management (2018)
9. Ko, H., Witherell, P.: Machine learning-based continuous knowledge engineering for additive manufacturing. IEEE (2019)
10. Deswal, S.: Modelling and parametric optimization of FDM 3D printing process using hybrid techniques for enhancing dimensional preciseness. Int. J. Int. Des. Manuf. (IJIDeM) **13**, 1197–1214 (2019)
11. Dey, A., Hoffman, D., Yodo, N.: Optimizing multiple process parameters in fused deposition modeling with particle swarm optimization. Int. J. Interact. Des. Manuf. (IJIDeM) **14**(2), 393–405 (2019). https://doi.org/10.1007/s12008-019-00637-9
12. Kolad, A., Dalu, R.: Optimization of process parameters in fused deposition modelling. Int. J. Innovative Res. Sci. Eng. Technol. **6**(1), 505–511 (2017)
13. Huang, D.J., Li, H.: Review of machine learning applications in powder bed fusion technology for part production. Nayang Technological University (2018)

Rx-PETaL: A Reactive Data Publishing Framework

Carl Camilleri(✉), Joseph G. Vella, and Vitezslav Nezval

Computer Information Systems, University of Malta, Msida, Malta
{carl.camilleri.04,joseph.g.vella,vitezslav.nezval}@um.edu.mt

Abstract. Organisations rely on the publishing of internal datasets to achieve crucial objectives, such as the implementation of an open data policy. Rx-PETaL is a framework that leverages concepts of reactive streaming and the Actor model to design efficient ETL processes for data publishers. We describe Rx-PETaL and illustrate its applicability to a number of real-world problems, as well as the results of the empirical evaluation of one application of Rx-PETaL, improving runtime duration of a legacy sequentially-processing application by more than 95%.

Keywords: Open data publishing · Reactive · Streams · Actor model

1 Introduction

Publishing internal datasets to various consumers facilitates business priorities, such as publishing of products in marketplaces, submission of online URLs to search engine crawlers, and publishing of data in standard and open data formats, which is a crucial enabler for the implementation of an "Open Government" [4] and Open Urban Platforms [15]. "Open" data is characterised by eight principles [4], namely that it must be complete, extracted from the primary data source, timely, accessible, machine-readable, non-discriminatory, published in a non-proprietary format, and license-free.

Some of these criteria apply to any type of data published by organisations. For example, data pushed to an aggregator website by an e-Commerce system should come from the original source, be timely and machine-readable to ensure optimal time-to-market, whilst preserving relevance even for time-sensitive information such as stock levels and prices. Publishing processes can be modelled along the concepts of Extract-Transform-Load (ETL) processes, where datasets stored in operational systems are read, integrated and transformed to a suitable format to be loaded in Descision Support Systems (DSSs) for ad hoc reports. Traditionally, ETL processes a) load data in a data warehouse; b) are *pull-based*, where the owner of the data warehouse is responsible to pull data from various data sources [18]; and c) run on a dedicated platform and whose coordination of import is set by the data warehouse infrastructure operations. *Publish ETL* (p-ETL), in contrast a) publishes datasets to different systems, even those unrelated to DSS; b) generates data in various formats, including open data formats such

L. Garg et al. (Eds.): ISMS 2020, LNNS 303, pp. 93–104, 2022.
https://doi.org/10.1007/978-3-030-86223-7_10

as the Resource Description Framework (RDF); c) is *push-based*, where the owner of the data source is responsible to push data to the different consumers; and d) timeliness of execution is important. For frequently-changing datasets where the data consumer expects relevant data, p-ETL processes execute frequently and so, duration and efficiency are two key requirements [13]. Duration is the time it takes for a p-ETL process to publish a fresh dataset. Efficiency defines the infrastructure resources needed to run the p-ETL processes, determining if the process can run on commodity hardware without disrupting operational transaction processing, which is a crucial quality-of-service (QOS) demand [16].

In this paper, we tackle the problem of designing fast, efficient and scalable p-ETL processes that run on commodity hardware and which are capable of generating complete and machine-readable datasets that are extracted from the primary data source. This problem has been studied in the literature and shown to be used in practice [12], but our novel approach considers the concept of reactive systems and the Actor model of computation for the implementation of p-ETL processes. We propose the Rx-PETaL framework, apply it to the design and implementation of real-world p-ETL business processes, and empirically show the advantages of our approach by improving the computation duration by 95% compared to a legacy sequential p-ETL process.

2 Literature Review

2.1 Background

Reactive systems perform computation on the occurrence of identifiable events [17] and are Responsive, Resilient, Elastic and Message Driven[1]. The *Observer* design pattern [8] defines *Observable* or *Subject* entities, to which *Observer* entities are subscribed. Observers are notified when an Observable's state changes.

Data Streams specialise this design pattern and express logic in terms of a series of steps that transform a flow of data [5]. Streams comprise of a *Publisher*, the source of the data, and a *Subscriber* that consumes the data, possibly linked with a number of *Processors* that transform the data as it flows through the stream. A stream can be *push-based*, where the Publisher is responsible to push data to the downstream stages, or *pull-based*, where the Subscriber is responsible to request data from upstream stages. The stream source can be *infinite* (e.g. IoT applications where the publisher is a sensor generating events continuously), or *finite* (e.g. a publisher that reads a data file and emits each record as a single data element in the stream).

Reactive streams[2] introduce the concept of non-blocking *back pressure* in pull-based data streams [5,6] and express asynchronous interactions between the different parts of the stream such that switching execution context is done at any time. Back pressure strategies are defined to deal with fast publisher/slow

[1] The Reactive Manifesto: https://www.reactivemanifesto.org/.

[2] The Reactive Streams specification: http://reactive-streams.org.

consumer situations and thus, reactive streams are specifically defined to deal with handling large streams of data.

Data Chunking is the process of grouping data elements in batches of a pre-defined size or within a pre-defined time window. Each batch, or *data chunk*, is a group containing at least one data element from the source dataset, in a way that processing a data chunk yields the same result as if the data elements within the chunk are processed individually. The idea of data chunking has been put forward to segment a dataset which is too large to fit in memory [13].

2.2 Related Work

A number of studies discuss data processing pipelines to publish finite but potentially-large datasets. *Triplify* [3] publishes relational data to various formats, including RDF triples, Linked Data, JSON and CSV. It aims to lower the technical and knowledge barrier for content owners to publish their datasets in machine-readable formats. The importance of the approach is shown through its application to a number of use cases, namely a) the product catalogue of an e-commerce merchant; b) the contents of an online blog; c) the dataset of an online image gallery; and d) the dataset from the OpenStreetMap project[3]. *ShExML* [9] is a generic language capable of encoding the transformation of heterogeneous datasets to RDF based on Shape Expressions. The approach focuses on lowering the technical barriers and development costs for RDF publishing pipelines, and a survey of 20 participants showed it to be simpler and more efficient to use, compared to SPARQL-Generate and YARRRML. *LinkedPipes ETL* (LP-ETL) [14] has similar objectives for streamlining the setup of workflows that lift internal datasets and publish them as linked open data. LP-ETL is applied to diverse use cases, especially in government organisations implementing open government initiatives [12], including the publishing of a) an internal dataset comprising measurements of temperatures in the municipality of Brno; and b) the national data catalogue from proprietary XML APIs in the Ministry of the Interior of the Czech Republic, for consumption by the European Data Portal. *DBPedia* EF [2], subsequently re-defined as a set of declarative data transformations in a generic process pipeline tool [7], extracts information from Wikipedia[4] and publishes it as RDF triples. *Data preparation* [11] is compared to traditional ETL, where the importance of establishing flexible but well-defined frameworks that assist a wide spectrum of users in the fusion of heterogeneous datasets to solve diverse problems is highlighted. *Chamaleon* [19] proposes a framework and a domain-specific language (DSL) to describe data wrangling pipelines.

3 The Rx-PETaL Framework

Rx-PETaL is a set of guidelines that we propose for the design and implementation of efficient p-ETL processes, based on the principles of reactive systems.

[3] http://www.openstreetmap.org.
[4] https://www.wikipedia.org/.

The framework is implementation-agnostic but we assume that the language or environment used to implement a system designed according to Rx-PETaL, supports the Actor model and Reactive streams.

3.1 Dataset Scoping

In the Dataset Scoping phase, the source dataset(s) and target dataset are identified. Furthermore, the source dataset(s) need to be classified in terms of their relation to the target dataset. Four types of source datasets are identified.

The **Lookup** datasets are frequently accessed during the publishing process, and are small enough to cache in main memory as a data structure that allows efficient (possibly $O(1)$) access during processing.

The **Primary** dataset defines the size of the target dataset: the output is *complete* if it contains all the necessary data items from the Primary dataset. An example is the list of products that constitute the catalogue to be exported to an aggregator website. The Primary dataset is designated as the *stream source.*

The **Linked** datasets are fragments of data which are linked to one or more data items within the Primary dataset. In contrast to Lookup datasets, they are typically too large to fit in memory. Taking the same example, a linked dataset for a Primary dataset of products is the list of media associated to each product.

The **Dependant** datasets constitute data which is extracted after analysing the whole database. For example, and keeping to the product catalogue use case, if the target dataset requires a numeric index for each product depending on the levels of stock available, an operation that traverses the whole source database is needed in order to generate this index. Such operations precede actual streaming, in order to generate a dataset which can either form part of the Primary dataset, or be classified as either Lookup or Linked.

3.2 Reactive Stream Design

This second phase considers the design of the Reactive stream. Our definitions are based on phrasing compatible with the popular Akka toolkit[5], also adopted in literature on reactive streaming [10]. Where possible, we give equivalent terminology used in ReactiveX[6], another common implementation of reactive streams.

Firstly, the *source* and *sink* of the stream need to be defined. From the Dataset Scoping phase, the Primary dataset is identified as the stream source, or a cold Observable. The stream's sink, or the subscriber of the reactive stream, is typically a null terminator operator.

Subsequently, the logic for the transformation and loading phases of the p-ETL process are defined as stream flows, or processors. The flows are designed and implemented specifically for the data process however, in Rx-PETaL we propose guidelines for the design of an efficient p-ETL process. For each guideline, the reactive stream operator supporting its implementation is indicated.

[5] https://doc.akka.io/docs/akka/current/stream/index.html.
[6] http://reactivex.io/.

Data Buffering. Consider Buffering, or pre-fetching, when connecting two stream flows where the publisher can be slower than the subscriber. For example, in a typical p-ETL process, the stream's source will depend on some I/O (e.g. reading data from disk or from the network). In Akka Streams, the *Buffer* is a stateful operator which, when the downstream flow backpressures, collects data elements from an upstream flow up to a maximum number, before backpressuring. *onBackpressureBuffer* is an equivalent operator in ReactiveX.

Data Chunking. A data chunking operator should be introduced within the stream prior to a flow step that needs to access element-dependant data from a source that is more efficient at performing operations on sets of elements rather than on single elements. For example, if each streamed data element needs to be supplemented with other data coming from a relational database management system (RDBMS), data chunking should be introduced to collect a pre-defined amount of stream elements and fire one query to the RDBMS that collects data for all the elements in a data chunk. The *groupedWithin* operator in Akka Streams performs data chunking by transforming upstream elements into groups such that, given that the upstream source is producing elements, a group of elements is emitted at least once within a pre-defined time window, but each group is guaranteed to contain at least one and at maximum X data elements, where X is defined at implementation stage. The equivalent operator available in ReactiveX is the *Buffer* operator.

Parallelism and Asynchrony. CPU-bound and I/O-bound flows should be considered for parallelism. Assuming dedicated infrastructure, stream operations (such as transformations of data elements to a string representation) can be parallelised up to the amount of CPUs available. I/O-bound flows, such as those that connect to external systems (e.g. databases and APIs), should be asynchronous. The *mapAsync* operator in Akka Streams, and *Parallel flows* in ReactiveX, implement parallelism of specific stream flows.

3.3 Actor System Design

The design of an Actor system is optional, however it should be considered when designing processing that a) have the possibility of accessing critical resources concurrently or b) need to scale out to more than one machine. Critical resources are defined as artefacts for which access needs to be controlled, such as writing to a file on disk, or calling an HTTP endpoint.

In Rx-PETaL, critical resources are modelled as Actors. Access to a resource is never done directly from the reactive stream, but some flows send messages to the respective actors handling a critical resource. Thus, the flow within the reactive stream is modelled as a producer of messages, sent to an actor's mailbox for processing. By implementing message passing in a request/response manner within the stream's flow (in Akka terminology, using the *Ask*, rather than the *Tell* constructs) within the stream's flow, the reactive stream's back-pressure mechanism ensure that the actor's mailbox does not overflow.

A suitable Actor system can also be defined to scale out CPU-bound tasks across more than one machine. In-line with the parallelism objective in

Rx-PETaL's stream design guidelines, CPU-bound tasks can be parallelised and delegated to an Actor system running on a cluster of machines, effectively scaling out parts of the processing on clusters of commodity hardware.

A number of languages, frameworks and toolkits support the implementation of an Actor system that complements a reactive streaming p-ETL, including Akka, Akka.NET, Orleans, Cloud Haskell, Actix and Pykka[7].

4 Practical Applications of Rx-PETaL

Rx-PETaL has been implemented in a number of real-world use cases. A high-level overview of two such applications in the e-Commerce industry are illustrated in Fig. 1. Specifically, Fig. 1a illustrates a process that loads an RDBMS with a dataset of popular user search queries, mapped to a representative image of a relevant product from the publisher's product catalogue, with relevancy scoring obtained by interrogating an internal search engine via its REST API. Figure 1b shows a stream that maps the publisher's product catalogue to a consumer-specific CSV file format, embellishing the source dataset with further data from a linked dataset as part of the process.

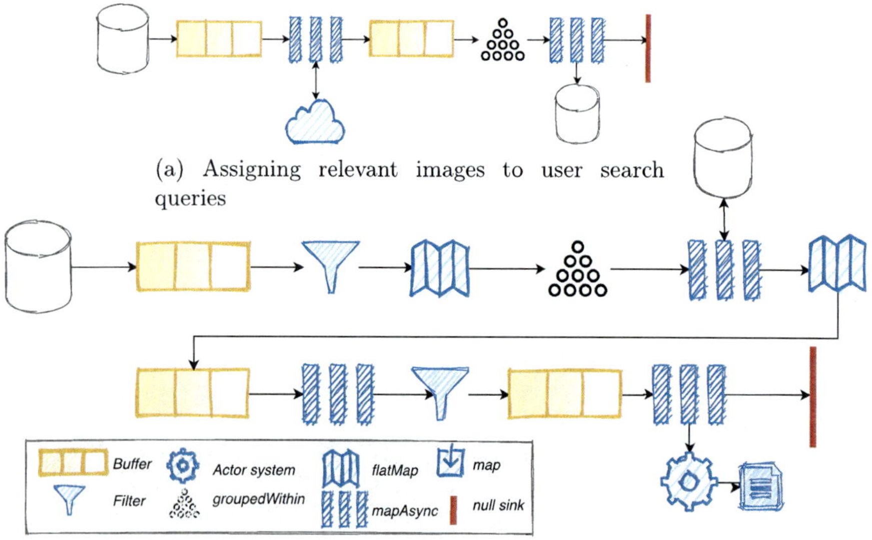

(a) Assigning relevant images to user search queries

(b) Publishing a product catalogue to CSV files

Fig. 1. Applications of Rx-PETaL to various publishing p-ETL processes

[7] https://akka.io/, https://getakka.net/, https://dotnet.github.io/orleans, https://dotnet.github.io/orleans, https://haskell-distributed.github.io/, https://actix.rs/, https://www.pykka.org/.

Throughout the rest of this section, we analyse in more detail a third application of Rx-PETaL, showing the design and implementation phases, as well as evaluating the performance of the resultant reactive streaming p-ETL process.

4.1 Problem Definition

We consider the problem of generating sitemap files for an e-Commerce merchant. The Sitemap XML protocol [1] defines a schema for XML files that are readable by automatic agents, known as "bots" or "crawlers". The contents of such an XML file allows these agents to discover a list of URLs that are served on a particular URLs. This allows search engine crawlers to discover these sitemap files, and is one of the ways through which a website administrator ensures that all the important URLs are discovered and included in a search engine index. Sitemap files can be published in both raw and compressed formats, and each file cannot contain more than 50,000 URLs, or be larger than 50 MB in size [1]. When the content exceeds any of these limits, the sitemap spans several files, each file respecting these limits, with an additional *sitemap index file* published.

The source dataset is stored in a Microsoft SQL Server 2016 RDBMS. The multi-tenant system hosts approximately 1.1 million products, 29 tenants and 33 locales. Most products are published in all locales, with a few exceptions. Each record in a table *Products* contains a comma-separated list of *TenantIds* in which it is available. The system therefore needs to generate 22 sitemap files and 1 sitemap index file per available locale.

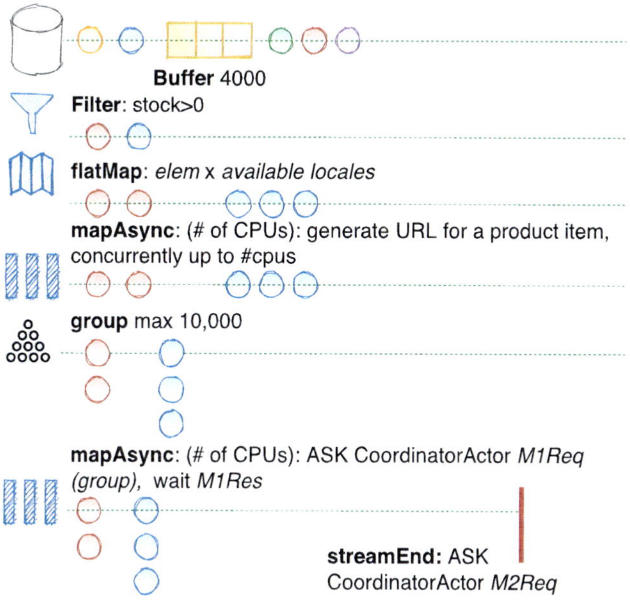

Fig. 2. Marble diagram of the reactive stream to generate Sitemap XML files

4.2 Design and Implementation

We follow Rx-PETaL guidelines to design and implement a reactive streaming problem for the Sitemap generation problem domain.

Dataset Scoping: Two types of datasets are identified. The table *Locales* is classified as a *Lookup* dataset because a) it contains merely 33 records and is not envisaged to grow beyond what fits in memory; and b) the information from this dataset is heavily used to generate URLs exported in the sitemap file. The table *Products* is classified as the *Primary* since the p-ETL process needs to generate at most one sitemap file entry for every record in this table.

Reactive Stream Design: Figure 2 depicts the reactive stream designed for this process. The notation in Fig. 2 roughly follows the convention of a marble diagram[8] with the stream's *source*, *flows* and *sink* on each line.

Actor System Design: Sitemap files stored on disk are not considered thread-safe and, consequently, are identified as resources that should be protected from concurrent write access. Furthermore, we identify the need to keep track of the number of URLs published in any one sitemap file, in order to ensure that not more than 50,000 URLs are included. Figure 3 illustrates the Actor system designed for this process.

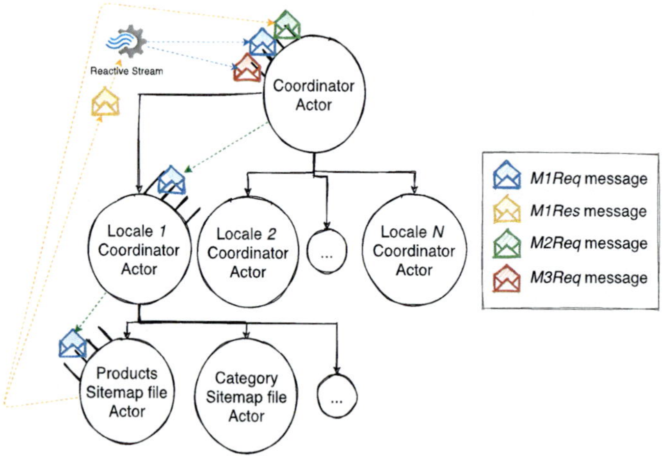

Fig. 3. Actor system for sitemap XML generation

The Reactive stream sends messages of type *M1Req* to a single *Coordinator Actor*. The message is a triplet of $\{locale, type, urls\}$. The *Coordinator Actor* supervises a set of child *Locale* actors, one per locale, and *forwards* any message received to the locale's corresponding actor. The *Locale Actor* actor supervises

[8] https://rxmarbles.com/.

a set of child *Sitemap File* actors, one per type of sitemap being generated. Finally, the *Sitemap File* actor is responsible to maintain in its state: a) the last sitemap XML file created and b) a counter C, initialised at 0 and indicating the number of *url* elements already written in the physical file. Upon receipt of an *M1Req* message, the *Sitemap File* actor performs the logic illustrated in Fig. 4. The *Coordinator Actor* is also responsible to maintain a list l of sitemap files per locale. Upon receipt of a message of type *M2Req*, the *Coordinator Actor* adds the details of the file to its state. Finally, upon receipt of an *M3Req* message, the *Coordinator Actor* persists l as one sitemap index file per locale.

```
//Actor initial state:
f <- nil //the name of the file currently being used
nf <- 1 //the file number
c <- 0 //the number of URLs written to the current file
t <- "products" //the type of sitemap files being generated
func processMessage(m:M1Req)
     IF f = nil OR |m.urls| + c > 50000
        f <- t_nf.xml //filename format: products_1.xml
        nf++
        Send message M2Req(m.locale, f) to Coordinator Actor
     END
     Write m.urls to f
     c += |m.urls|
     Send message M1Res to reactive stream
```

Fig. 4. Processing *M1Req* message by the Products Sitemap File Actor

Implementation: We implemented the reactive stream and actor system in C# on top of .NET Core 5, and using the Akka.NET toolkit that supports both reactive streams and the Actor model.

4.3 Evaluation

A legacy process existed that followed a non-reactive approach: each record from the Primary dataset was loaded from the RDBMS, filtered as necessary, and transformed to its XML representation to generate the sitemap files. The legacy process was sequential, in that it processed the primary dataset one record at a time. It was implemented in C#, on top of .NET 4.8, and hosted on 1 virtual machine having 5 CPUs @2.30 GHz and 10 GB of RAM.

We deployed the reactive p-ETL process on the same infrastructure, connected to the same database, and executed 10 sequential runs of each process. Figure 5 illustrates the duration of both processes across each run. During each run, 2.7 GB of data was generated across 477 XML files, comprising 17,043,558 sitemap entries. For the legacy process, the minimum duration observed was 2 hr 25 min, with a maximum peak of 4 hr 41 min and an average of 2 hr 43 min. The peak is explained by the fact that the Sitemap XML generation process performs heavy queries on the source RDBMS to read the whole dataset, and consequently competes with other processes for hardware resources of the RDBMS server. In contrast, the maximum execution time of the streaming p-ETL process observed was 9 min 29 s, whilst the minimum duration was 7 min 20 s, for an average duration of 7 min 34 s. It was also observed that the legacy process required 700 MB of RAM to execute, compared to 230 MB used by the streaming process, and 10% of the available CPUs, compared to the 80% CPU usage of the streaming process.

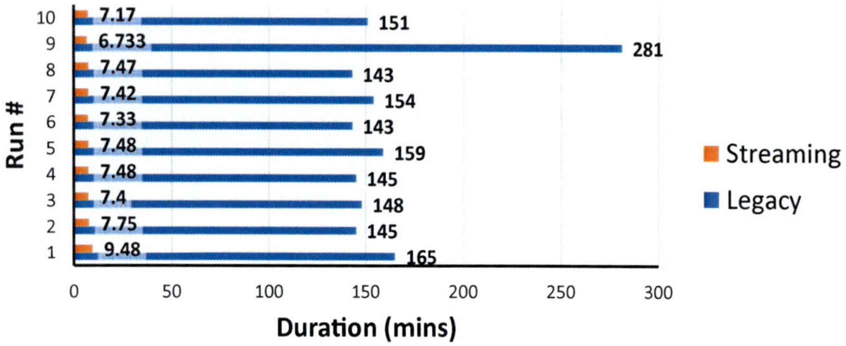

Fig. 5. Duration of the legacy *vs.* the streaming p-ETL process

5 Conclusion

Rx-PETaL is a novel approach to the design of efficient p-ETL processes. The framework provides guidelines for the design and implementation of such processes, based on the principles of reactive systems. We provide details of reactive stream operators that are suitable to process different types of data sources. The aspect of concurrent access to critical resources, such as file system resources, is also tackled, and we give guidelines on the use of the Actor model to perform efficient access to such resources.

Compared to other works in the literature, our framework has been shown to be applicable to diverse use cases. We do not consider the publishing of datasets only in semantic web formats, such as RDF, but also other proprietary formats and protocols, such as the XML Sitemap protocol, adopted by systems, such as search engines, which a dataset owner may need to integrate with.

We also focused on performance and efficiency, and showed that the runtime duration of an implementation of Rx-PETaL that generates more than 17 million records finishes in less than 10 min when running on commodity hardware, an improvement of 95% compared to the duration of a legacy process. Other works in the literature place less focus on this aspect of publishing processes, or are unsuitable for large datasets due to long execution times.

To the best of our knowledge, Rx-PETaL is unique in uplifting reactive streaming and Actor modelling to achieve the necessary duration, efficiency and client application QOS properties required by frequently-running p-ETL. Rx-PETaL is currently a framework, but it is envisaged to be implemented as a system and a library of pluggable stream processors, making it available as a self-service data wrangling tool, with a UI to build and debug p-ETL workflows.

Acknowledgement. This work is partly funded by the ENDEAVOUR Scholarship Scheme (Malta), part-financed by the European Union – European Social Fund (ESF) under Operational Programme II – Cohesion Policy 2014–2020.

References

1. Sitemaps XML. https://www.sitemaps.org/protocol.html. Accessed 10 Sept 2020
2. Auer, S., Bizer, C., Kobilarov, G., Lehmann, J., Cyganiak, R., Ives, Z.: Dbpedia: a nucleus for a web of open data. In: Aberer, K., et al. (eds.) The semantic web, vol. 4825, pp. 722–735. Springer, Heidelberg (2007). https://doi.org/10.1007/978-3-540-76298-0_52
3. Auer, S., Dietzold, S., Lehmann, J., Hellmann, S., Aumueller, D.: Triplify: light-weight linked data publication from relational databases. In: Proceedings of the 18th International Conference on World Wide Web, pp. 621–630 (2009)
4. Bauer, F., Kaltenböck, M.: Linked open data: the essentials. Edition mono/monochrom, Vienna, vol. 710 (2011)
5. Calia, A., et al.: Streaming pool-managing long-living reactive streams for Java. In: ICALEPCS2017, Barcelona, Spain (2017)
6. Dadel, P., Zieliński, K.: Evolution of reactive streams API for context-aware mobile applications. Comput. Inform. **35**(4), 852–869 (2017)
7. De Meester, B., Maroy, W., Dimou, A., Verborgh, R., Mannens, E.: Declarative data transformations for Linked Data generation: the case of DBpedia. In: Blomqvist, E., Maynard, D., Gangemi, A., Hoekstra, R., Hitzler, P., Hartig, O. (eds) European Semantic Web Conference, pp. 33–48. Springer, Cham (2017). https://doi.org/10.1007/978-3-319-58451-5_3
8. Gamma, E.: Design Patterns: Elements of Reusable Object-Oriented Software. Pearson Education India (1995)
9. García-González, H., Boneva, I., Staworko, S., Labra-Gayo, J.E., Lovelle, J.M.C.: ShExML: improving the usability of heterogeneous data mapping languages for first-time users. Peer J Comput. Sci. **6**, e318 (2020)
10. Ge, X., Zhu, H., Wang, C., Yuan, Z., Zhu, Y., et al.: Design and implementation of reactive distributed Internet of Things platform based on actor model. In: 2019 IEEE 3rd Information Technology, Networking, Electronic and Automation Control Conference (ITNEC), pp. 1993–1996. IEEE (2019)
11. Hellerstein, J.M., Heer, J., Kandel, S.: Self-service data preparation: research to practice. IEEE Data Eng. Bull. **41**(2), 23–34 (2018)
12. Klímek, J., Škoda, P.: LinkedPipes ETL in use: practical publication and consumption of linked data. In: Proceedings of the 19th International Conference on Information Integration and Web-based Applications & Services, pp. 441–445 (2017)
13. Klímek, J., Škoda, P.: Speeding up publication of Linked Data using data chunking in LinkedPipes ETL. In: Panetto, H., et al. (eds.) OTM Confederated International Conferences "On the Move to Meaningful Internet Systems". pp. 144–160. Springer, Cham (2017). https://doi.org/10.1007/978-3-319-69459-7_10

14. Klímek, J., Škoda, P., Nečaský, M.: LinkedPipes ETL: Evolved linked data preparation. In: European Semantic Web Conference, pp. 95–100. Springer, Cham (2016). https://doi.org/10.1007/978-3-319-47602-5_20
15. Lämmel, P., et al.: Metadata harvesting and quality assurance within open urban platforms. J. Data Inf. Qual. (JDIQ) **12**(4), 1–20 (2020)
16. Lu, C., Alvarez, G.A., Wilkes, J.: Aqueduct: online data migration with performance guarantees. In: FAST, vol. 2, p. 21 (2002)
17. Salvaneschi, G., Margara, A., Tamburrelli, G.: Reactive programming: a walkthrough. In: 2015 IEEE/ACM 37th IEEE International Conference on Software Engineering, vol. 2, pp. 953–954. IEEE (2015)
18. Sreemathy, J., Joseph, V., Nisha, S., Chaaru Prabha, I., Gokula Priya, R.M.G.: Data integration in ETL using TALEND. In: 2020 6th International Conference on Advanced Computing and Communication Systems (ICACCS), pp. 1444–1448 (2020)
19. Valencia-Parra, Á., Varela-Vaca, Á.J., López, M.T.G., Ceravolo, P.: CHAMALEON: framework to improve data wrangling with complex data. In: Krcmar, H., Fedorowicz, J., Boh, W.F., Leimeister, J.M., Wattal, S. (eds.) Proceedings of the 40th International Conference on Information Systems, ICIS 2019, Munich, Germany, 15–18 December 2019. Association for Information Systems (2019). https://aisel.aisnet.org/icis2019/data_science/data_science/16

Design and Implement Smart Home Appliances Controller Using IOT

Bhupesh B. Lonkar[1](✉), Annaji Kuthe[2], Ritesh Shrivastava[3], and Pallavi Charde[4]

[1] Datta Meghe Institute of Engineering, Technology and Research, Wardha, India
[2] Karmavir Dadasaheb Kannamwar College of Engineering, Nagpur, India
[3] Anjuman College of Engineering, Nagpur, India
[4] G. H. Raisoni College of Engineering, Nagpur, India

Abstract. Today, most of the appliances are working on the internet platform to use different sensors and other equipment. The devices can communicate with each other when it allow to the data over the network. This technique is implemented in our home to handle appliances automatically and easy. The source of the current task is performed by the home appliance controller inside the house. Now a day's, it is very easy to develop the home appliance controller, because high computational devices, user friendly and cost effective automation system. For the communication, different technology of it using by home appliance controller like Wi-Fi, Bluetooth and controlling varies appliances it used different devices like smart phone, tablet and laptop. The existing systems may not effective for general user and it causes trouble to user due to use by the web application for looking the view of the house. If the users want to see the status of the home, then each and every time he/she can access the web application. Therefore, the purpose of development of this system is to let people aware about these technologies, and make the system as simple as possible for an general peoples to understand. We are going to mobile application for controlling the home appliance form any remote location using internet of things.

Keywords: MOSFET · IOT · WI-FI · TTL · IC · CMOS · IDE · SQL

1 Introduction

Today, Due to high demand of automated systems so that it reduces human intervention. For finding vertical domain on the Internet of Things (IOT) platform, it may use a set of new applications and their services. The smart home domain used market segment which is represented by home automation. According to survey of market research company Statista1, by the year 2020, value of Smart Home market worldwide will reach $ 43 Billion, which is three times more compared to that in the year 2014. As per the US Business Analyst Company, IHS2 evaluate that the emerging smart homes are using 470 smart appliances. With

L. Garg et al. (Eds.): ISMS 2020, LNNS 303, pp. 105–117, 2022.
https://doi.org/10.1007/978-3-030-86223-7_11

the development of modern homes, the home automation will play an important role [4].

The home appliance controller is the modern technology which is very useful for handicapped people. The home appliance controller is connected to the system which is easily controlled or handle by user from control devices. In this system, the user can be monitored and controlled the home appliances, and it interacts with system in a user friendly interface. With the help of Internet of Things (loT), the home automation development is becoming much simpler and popular [6].

This system relies on the internet of things to provide the features of the home appliance control to operate easily. It used android application to basically convert a Smartphone device into a remote device for home appliances. This system allows the user can automatically control switching on of lights, fan speed control, curtains. The user can use the internet for accessing the complete IoT system from anywhere side. But the internet connectivity must always have for ESP32 controller [10].

In this paper, we have mentioned the home appliance controller model to easily handle home appliances through a short range service like Bluetooth, WiFi on IOT. Also, its focus on a remote location in which the system can easily handle and controlled by the user.

The system is working on Android platform in which it is easily handled and controlled with human less effort and very less cost for any users.

2 System Components

2.1 ESP32

In Fig. 1 express If the system is created ESP32 a low-cost system-on-chip (SoC) series. ESP32 is the advance version of ESP8266 that is widely used in IoT system. The ESP32 has enabled with Wi-Fi and Bluetooth technologies, which make it for the development of IoT system and embedded systems [9].

Fig. 1. ESP 32 device

2.2 Relay

Relays are the switches that perform the operation of open and close circuits electromechanically or electronically. From Fig. 2 the relay is connected to ESP32 and the output of it connect as a sequence of (i) light (ii) fan (iii) curtains in home appliances. Relay required low current and voltage and activate the switch which is connected to a high voltage. The relay takes 5V supply enabled with 4 input pins which connects to ESP 32 and can trigger up to 10A, 250V supply [8].

Fig. 2. Relay channel

2.3 ULN2003

Fig. 3. ULN2003 device

Above Fig. 3 ULN2003 is a relay driver IC. It consist of a Darlington array having high voltages and high currents as well. It is powered with seven open collector Darlington pairs having a common emitter. ULN2003 has a compatible to handling seven different relays at a time. A single Darlington pair having two bipolar transistors and it's running on the current range of 500 mA to 600 mA. ULN2003 control on 5V and TTL (Transistor Logic) and CMOS (Complementary Metal Oxide Semi-conductor).The ULN2003 pin configuration is that the input pins are made up on the left side of the IC, whereas the output pins are made up on the right side in front of the corresponding input pin. This IC has a lot of applications. They are acting as relay drivers to run different types of loads [9].

ULN2003 Pin out:
ULN2003 has 16 input pins. Output pins as follows:

- Pin-1 to 7 = 7 Output pins
- Pin-10 to 16= 7 Output pins
- Pin-8= 1 Ground pin
- Pin-6= 1 COM pin

2.4 MOSFET

Fig. 4. MOSFET

The above Fig. 4 MOSFET (Metal Oxide Semiconductor Field Effect Transistor) transistor is a semiconductor device which converts and boost Subsequent paragraphs, however, are indented. electronic signals in the electronic devices. Due to small size, the MOSFET is a designed and makeup in a single chip. The MOSFET is connected to the source terminal so that it creates three terminal device like a field effect transistor. The MOSFET is transistor and it can be used commonly in both analog and digital circuits [11].

3 Architecture of Smart Home Appliance Control System

The system architecture is associated with the two part hardware and the software part. In this architecture Fig. 5 is showing the system, how it will going to be work. We are going to use cardboard to make our experimental setup that contains kits, power supply and devices. This device, control of the mobile application that is designed in the android application making software. The architecture of system gives a brief Idea of operation home appliance controller using IoT [13].

This system consist of four different appliances such as fan, light and curtains which are operated and controlled automatically using Wi-Fi and through an application installed on android. ESP32 is provide an open-source electronics platform for using hardware and software. For the widest range of communication, ESP32 connected to these appliances with its digital input/output ESP32 is a series of low-cost, low-power system on chip microcontrollers with integrated

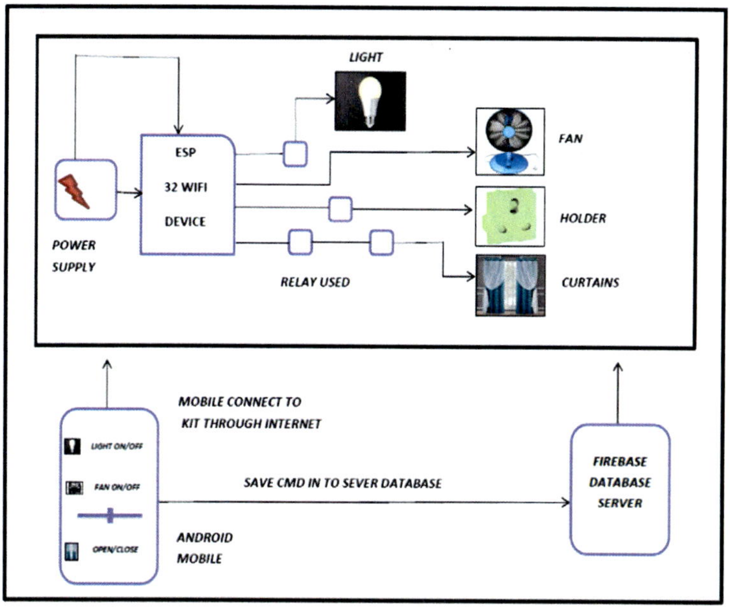

Fig. 5. System architecture

Wi-Fi and dual-mode Bluetooth. We have been using relays for ON/OFF operation mode and Google firebase for storing data. Relays act like a switches that open and close circuit electromechanically or electronically [7].

So we will firstly connect our hardware component to each other i.e. connection of ESP32 and all devices and relays for ON/OFF. For the driving the relays we used the ULN2003 IC that is relayed diver IC with connected between the relay and the ESP32 device. For the control fan we used MOSFET because we have given a speed controller seek bar into the android application. We have given an ON/OFF buttons for the light and fan and a speed adjustment bar for fan for controlling the speed of the fan. For curtains, we have one button for ON/OFF. This all hardware component is connected to the power supply, we used a Google firebase database server for storing our data and command for controlling our system. The system is using google firebase database server to store the information in real time on cloud hosted. There is no SQL database used inside the system. We have developed the application through android studio, which is starting on the platform of android for using the IDE (Integrated Development Environment) or tool (layman terms) [5].

Our application contains a user friendly interface such that the user can easily use and access the feature of the system. Application have different buttons for different functions. It have ON/OFF button for each module, and a speed adjustment bar for controlling the speed of the fan. And our app is also going to notify that if the fan is on or off or curtain is closed or open. Our Google firebase database server is connected to hardware component as well as our software i.e. application [12].

4 Methodology

The Appliances will operate using android mobile device. It is informed regarding the status of development/ design of the "Home Appliance controller using IOT". The involvement of things that is shown in below flow chart

1. Switch ON/OFF Light
2. Switch ON/OFF Fan with Speed Controller
3. Curtains Open/Close

Below Fig. 6 is of used methodology. In this methodology the user will first give the power supply to the all components as well as kits that we are going to connect to our system. Then all users connect the mobile device with the ESP 32 Kit with WIFI connection and establish the connection.

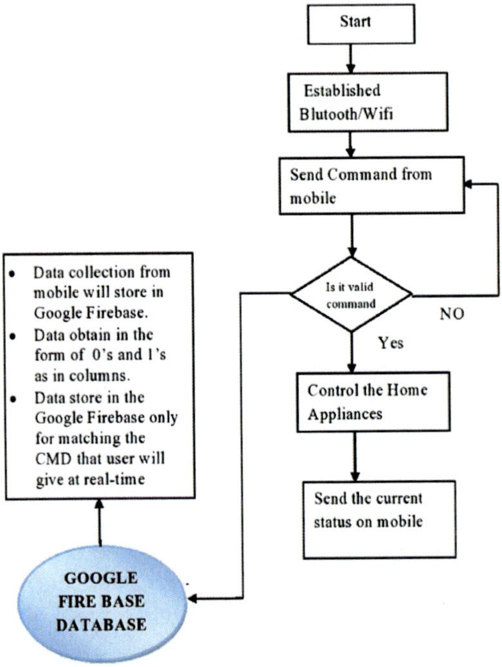

Fig. 6. Methodology flowchart of system

After Establishing connection user will give the command through the mobile and it will generate results of the component. Whenever we will give command it will save into the Google Firebase Database. And it will check the CMD status, we will going to give through mobile that are equal or not in the database.

This system can be implemented using low cost electronics and software technology, built It reasonable, technically and workable. This system used the android device platform and a few electronic components like ESP32, relay switches, etc. This system is deploying the wireless technology and embedded systems which are fairly with currently used technology. Therefore, it is very much recommended by the technology.

5 Working Mode

5.1 Actual Hardware Model

Figure 7 below is the hardware model of our project. It contains cardboard having all devices on it. Like Light Holder, Fan, Curtain Setup, extra holder, PCB sheet all electronics devices. It's all connected to each other via wires. It's also having a SMPS to provide power supply to all devices. And one plug for main power supply.

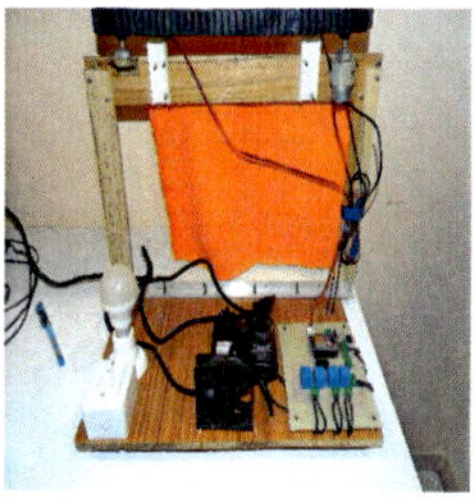

Fig. 7. Hardware model

5.2 Light Mode

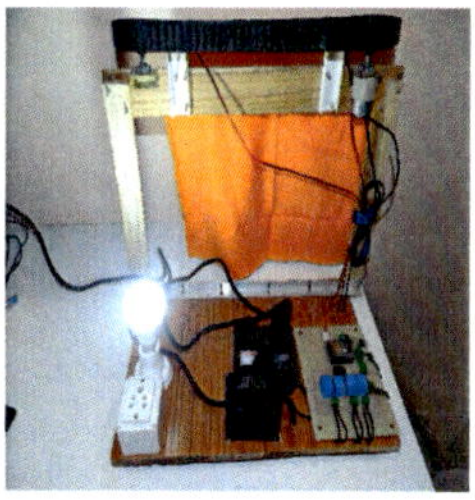

Fig. 8. Light ON module

Figure 8 below is the hardware model of our project. It contains cardboard having all devices on it. But above Fig. 8 show the one device is in ON mode that is light that means the user is controlling the device via mobile phone, as well as Bluetooth.

5.3 Curtain ON/OFF Module

It contains cardboard having all devices on it. But Fig. 9 show the one device are in ON mode that is Curtain that means curtain is open the user is controlling the device via mobile phone, As well as Bluetooth. Also Fig. 10 is showing the Curtain module that curtain is close. Mean the devices in off mode.

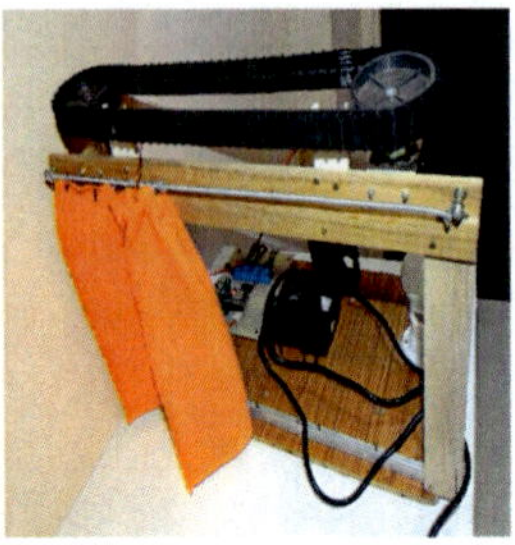

Fig. 9. Curtain ON module

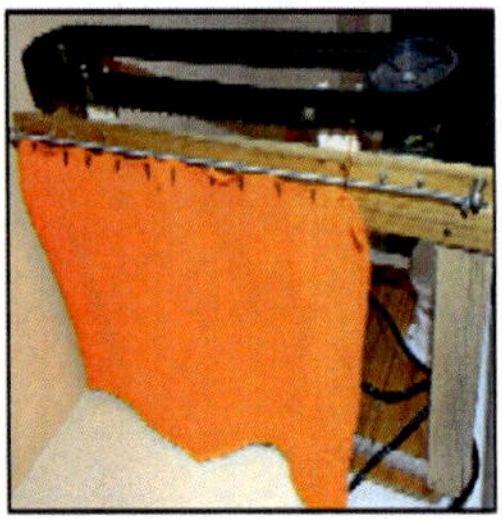

Fig. 10. Curtain OFF module

5.4 Fan ON Module

Figure 11 below is showing the one device is in ON mode that is Fan that means curtain is ON mode the user is controlling the device via mobile phone, as well as Bluetooth.

Fig. 11. Fan ON module

6 Result

A. Splash Screen

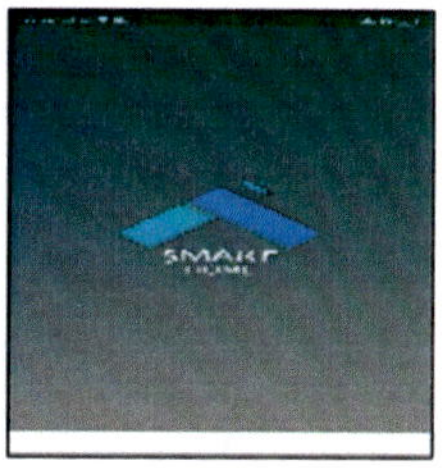

Fig. 12. Splash screen

Above Fig. 12 is flash screen of our mobile application which is shown at the beginning of the application. When application Start then this screen will appears for 5 s after few second it will vanish. It will show only one time whenever we will install the application for the first time.

B. Information Screen

Fig. 13. Information screen

Figure 13 above is an information screen of our mobile application which is shown at the beginning of the application. When application Start, it will show after splash screen this screen have next and skip button go to the next screen. After skipping it will directly go to the Login Screen and click on next button it will go to the next screen.

C. **Permission Screen**

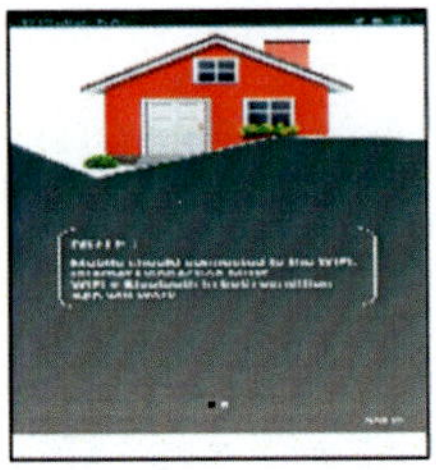

Fig. 14. Permission screen

Above Fig. 14 is a Permission screen of our mobile application which will show after the information screen, it have Got It button to go to the next screen. In this screen it will ask for Bluetooth permission to access the devices using the Bluetooth menu in the dashboard of our application.

D. **Login Screen**
Below Fig. 15 is the Login Screen it contains Edit text for username and password, and one button for the login. After clicking on Login button it will check id and password that was already in a fix in the code, if username and password is correct it will go to the Main Dashboard and show massage Welcome to the Home Appliances Controller otherwise it will show error message. Every user has same login id and password that is fixed for all.

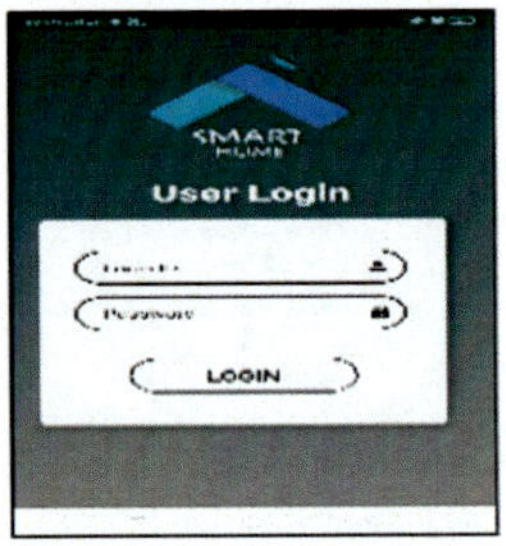

Fig. 15. Login screen

E. Dashboard using with Internet (OFF Mode)

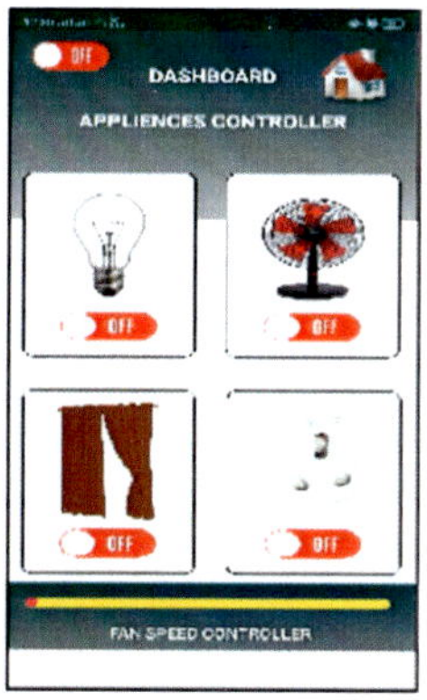

Fig. 16. Dashboard with using the internet (OFF Mode)

Above Fig. 16 is the Main Dashboard of our android application. Having some controller module Like the Light, Fan, Curtain, Holder and Fan Speed Controller seek bar. Every module/component having Toggle Button having functioned of ON and OFF. The dashboard also having one Toggle Button for Bluetooth. That can control devices via Bluetooth option. When Bluetooth Button OFF this dashboard controller work on Internet connection. This fig showing all controllers is OFF means all devices are now in off mode.

F. Dashboard using with Internet (ON Mode)

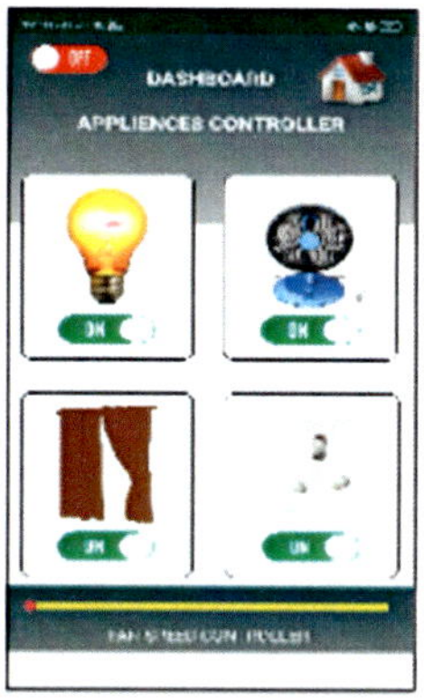

Fig. 17. Dashboard using with internet (ON Mode)

Above Fig. 17 is the Main Dashboard of our android application. Having some controller module Like The Light, Fan, Curtain, Holder and Fan Speed Controller seek bar. Every module/component having Toggle Button having functioned of ON and OFF. The dashboard also having one Toggle Button for Bluetooth. That can control devices via Bluetooth option. When Bluetooth Button OFF this dashboard controller work on Internet connection. This fig showing all controllers is ON means all devices are now in ON mode. Now this dashboard is working on the Internet.

7 Conclusion

Home automation and IOT is widely accepted technology across the world. Our goal is to provide useful system to the world. Using mobile the Presence of each and every module has been reasoned out and placed very carefully. Hence the contributing to the best working unit for automation of electrical devices has been designed perfectly. Secondly, using highly advanced device like ESP32, Relay modules, other devices with the help of growing technology, the project has been successfully implemented with a great idea. Thus the project has been successfully designed and tested.

In this system, the home appliance controller is used as a low cost hardware and user friendly environment presenting for the users. Our system gives better performance than existing Bluetooth based conventional home automation systems, it comes up with a simple approach for home automation. This system is not only providing the service for elderly and handicapped people, but also to provide benefits to reduce the human cost and save energy with the help of an Android application.

This approach is also time saving, because lots app time taken by user that will reduced due to android application through that platform user only click on that button, then this instruction goes to database it will check that electronic device is on or off this kind of instruction are control through the app that they can take less time for executing the user instruction this is the main Moto of our home appliances controller project.

Conflict of Interest

The authors declare that they have no conflict of interest.

References

1. Iqbala, A., et al.: Interoperable internet-of-things platform for smart home system using web-of-objects and cloud. Sustain. Cities Soc. **38**, 636–646 (2018)
2. G. of India (GoI).: Smart cities government of India (GOI). http://smartcities.gov.in/ (2016)

3. Ransing, R.S., Rajput, M.: Smart home for elderly care, based on wireless sensor network. In: 2015 International Conference on Nascent Technologies in the Engineering Field (ICNTE). IEEE (2015)
4. Jabbar, W.A., Ismail, M., Nordin, R.: Evaluation of energy consumption in multipath OLSR routing in smart city applications. In: 2013 IEEE Malaysia International Conference on Communications (MICC), pp. 401–406 (2013)
5. Song, T., Li, R., Mei, B., Yu, J., Xing, X., Cheng, X.: A privacy preserving communication protocol for IoT applications in smart homes. IEEE Internet of Things J. **4**, 1844–1852 (2017)
6. Li, M., Lin, H.-J.: Design and implementation of smart home control systems based on wireless sensor networks and power line communications. IEEE Trans. Ind. Electron.**PP**(99), 1 (2014)
7. Song, T., Li, R., Mei, B., Yu, J., Xing, X., Cheng, X.: A privacy preserving communication protocol for IoT applications in smart homes. IEEE Internet of Things J. **4**, 1844–1852 (2017)
8. Lee, K.-M., Teng, W.-G., Hou, T.-W.: Point-n-Press: an intelligent universal remote control system for home appliances. IEEE Trans. Autom. Sci. Eng. **13**, 1308–1317 (2016)
9. Jabbar, W.A., Alsibai, M.H., Amran, N.S.S., Mahayadin, S.K.: Design and implementation of IoT-based automation system for smart home. In: International Symposium on Networks, Computers and Communications (ISNCC), pp. 1–6 (2018)
10. Han, D.-M., Lim, J.-H.: Design and implementation of smart home energy management systems based on zigbee. IEEE Trans. Consum. Electron. **56**, 1417–1425 (2010)
11. Wu, S., Rendall, J.B., Smith, M.J., Zhu, S., Xu, J., Wang, H., et al.: Survey on prediction algorithms in smart homes. IEEE Internet of Things J. **4**, 636–644 (2017)
12. Popli, S., Jha, R.K., Jain, S.: A survey on energy efficient narrowband internet of things (NBIoT): architecture, application and challenges. IEEE Access **7**, 16739–16776 (2019)
13. Acharjya, D.P., Geetha, M.K., Sanyal, S. (eds.): Internet of Things: Novel Advances and Envisioned Applications. SBD, vol. 25. Springer, Cham (2017). https://doi.org/10.1007/978-3-319-53472-5

Check for updates

A Systematic Review on the Deployment of Massive Multiple-Input-Multiple-Output (MIMO) in Next-Generation Wireless Systems: Challenges and Prospects

Folarin Olaloye, Sanjay Misra(✉), Emmanuel Adetiba, and Jonathan Oluranti

Center of ICT/ICE Research, Covenant University, Ota, Ogun, Nigeria
{joseph.olaloye,sanjay.misra,
emmanuel.adetiba}@covenantuniversity.edu.ng

Abstract. The fifth generation (5G) wireless network systems is also known as worldwide wireless web (W4) because based on laid-out expectations, it does not require limitations. Though 5G is not fully deployed at present, one major technology that will contribute immensely to its reality is massive MIMO. There are, however challenges associated with it which serve as goldmines for future work. This review is aimed at gaining understanding of the laid-out specifications of 5G networks, the operation of massive MIMO as a technology for 5G and the challenges to be addressed by future work to enable the effective deployment of the technology to future generation networks. Various searches in science direct, Google scholar and IEEE Xplore are carried out to obtain all relevant studies. Papers that were considered range from the years 2010 through 2017 to ensure that current information is retrieved. From initial search processes, a total of about 500 papers that were related to the subject matter were retrieved. Out of about 70 filtered studies from the total obtained, 15 were selected after detailed analysis. This review will help to enhance further research works in massive MIMO technology as it exposes the associated limitations as well as some of the related areas being subjected to current research and investigation.

Keywords: Massive MIMO · Fifth generation (5G) networks · Spectral efficiency · Energy efficiency

1 Introduction

There was a dramatic transition from 3rd to 4th generation, and even greater advances are currently taking place in LTE and 5G. The standardization of 5G by standards bodies will continue through the year 2020, with continuous enhancements during the next decade. However, some operators are exploiting 5G in its initial standards level (pre-standard) and networks based on this standard level are expected to be deployed in trial by early 2019 [1]. LTE will not be totally replaced by 5G, but the integration and co-existence of both technologies will be applicable in most networks most likely in the late-2020s.

L. Garg et al. (Eds.): ISMS 2020, LNNS 303, pp. 118–131, 2022.
https://doi.org/10.1007/978-3-030-86223-7_12

Many applications that will make 5G a strong reality are not only products of LTE and its advanced forms but are also bringing into limelight the Massive MIMO technology. For instance, with carrier aggregation, mobile operators have not only exploited the potentials of their available spectrum for capacity and performance augmentation, but the technology also supports entirely new functionalities, such as exploring unlicensed bands [2].

The progress of the future generation (5G) wireless communication networks is being fueled by the ever-growing demand for fast communications of high reliability and importantly improved quality of user experience. It has been generally accepted based on theoretical specifications that the 5G wireless communication system will attain a capacity a thousand fold that of the long-term evolution advanced (LTE-A) system for fourth generation [3]. The 5G system is also required to have spectral efficiency of 3 to 5 times that of the existing LTE-A system. This is consequently equivalent to a maximum data speed of 50Gbps for low mobility users [4]. Massive MIMO is an evolving technology which brings MIMO to a greater (orders-of-magnitude) scale compared to the latest advanced level currently deployed.

Massive MIMO involves system using antenna arrays that have hundreds of antennas, serving many tens of user terminals simultaneously and in the same frequency-time domain. The rationale behind Massive MIMO is to actualize all advantages of traditional MIMO on a very large scale [5]. Massive multiple-input multiple-output (MIMO) technology has successfully attracted the attention of a lot of researchers because of its promising potential of greatly enhancing the robustness, spectral efficiency and energy efficiency of mobile systems [6]. A typical massive MIMO system comprises of a transmitter and a receiver. 10 to 100 antennas are provided for both transmitter and receiver. A massive MIMO has the capacity to support future digital infrastructure require for Internet of people, Internet of things, with clouds and other network infrastructure.

The last decade has witnessed tremendous advancement in mobile and wireless communications. According to [7], wireless networks technologies like 3G and 4G emerged as a result of increasing demand for high quality resources to cater for multimedia data especially. However, these technologies still appear insufficient to meet the demands hence the need for improved technologies like 5G networks that extend the capabilities of the 4G networks. Precisely, 5G networks emerged to tackle the shortfalls of 4G networks especially the need for better data rate and capacity at reduced cost with massive inter-device connectivity [8]. One such technology that has been proven to cater the requirements of next generation wireless systems is the massive MIMO. This technology has appealed to many researchers due to its promising capability of greatly improving spectral efficiency, energy efficiency, and robustness of the system [9]. This paper therefore attempts to conduct a systematic review of this supporting technology on previous related works which can be of usefulness to researchers for further investigation into this area of research.

This introductory section is followed by a critical literature review of previous studies related to massive multiple-input multiple-output as well as on-going technological advancement to 5G standards. The rest of this paper is structured as follows: "Methods" section shows the detailed procedures of retrieving relevant information on the subject matter. "Results" section gives the obtained results, including those tabulated in the review. Some current research areas related to massive MIMO are also highlighted in this study. These sections precede a concise conclusion.

2 Background

According to [10], multiple-input multiple-output (MIMO) technology was designed to meet the growing demand of increased data speed, reduced transmission problems such as multipath fading, and spectral efficiency, i.e. to improve throughput and accommodate the growing capacity demand with the limited available frequency spectrum. One of the fundamental functions of MIMO technology is spatial multiplexing in which a signal at high rate is divided into many streams each at lower rate. Each stream is then transmitted differently from each antenna at the transmitter, arrive the antenna array at the receiver and are combined effectively to give a high multiplexing gain and consequently increased spectral efficiency [11]. Furthermore, one major problem encountered in MIMO systems is the hardware cost and complexity, due to the incorporation of Radio frequency chain circuitry for each antenna. A technique known as antenna-subset selection has already been proposed to curb this obstacle, in which the radio frequency chains are less in number than the actual antenna elements. However, designing an optimal algorithm for antenna selection to meet the Quality of Service (QoS) requirements and reliability of transmission has been an engineering challenge.

In [12], it was stated that mobile operators are currently in dire need of technologies that provide any or both improved spectral efficiency and exploration of under-utilized portions of the spectrum. The future 5G networks will deploy massive MIMO which will curb the limitations of MIMO 4×4 and 8×8 implemented in commercial devices. Such limitations include high susceptibility to noise especially in dense urban environments with inherent refraction and reflection situations.

[13] opined that this research area has become worthwhile especially as current developments are being made towards the emergence of 5G networks. In attempt to practically apply this technology, some crucial challenges were identified. For example, the increase in the number of antennas gives rise to the complications of precoding and detecting operations especially when the antennas at the transmitter are less than, or comparable in number to the antennas at the receiver. Also, it is challenging to embed many antennas into a limited-size device with minimal correlations.

[14] stated that massive MIMO technology has potentials of delivering very high data rates and channel reliability without the need of additional bandwidth or transmission power. The various channel models that are important for the analysis and evaluation of the performance of massive MIMO systems are mainly grouped into the correlation-based stochastic models (CBSMs) and geometry- based stochastic models (GBSMs).

[15] opined that massive MIMO is very useful in the realization of emerging broadband networks in the following areas:

- Capacity increase due to the principle of spatial multiplexing, and dramatic improvement in the energy efficiency based on the constructive addition of useful emitted wavefronts intended for the receiver terminals, and cancellation of the non-useful wavefronts.

- Deploying massive MIMO with low-power, low-cost hardware components. Hundreds of such components replace the relatively high-power, expensive components e.g. amplifiers, large cables, etc. in the traditional systems. This also contributes to the resilience of massive MIMO because if one or few antenna units fail, it does not cause the collapse of the entire system.

Based on the law of large numbers, massive MIMO helps to significantly reduce latency due to the multipath fading encountered by signals propagated [16].

Furthermore, the quality of service on a massive MIMO system is majorly dependent on the channel state information (CSI) accuracy. This information must be consistent both at the forward and reverse channels between the mobile user equipment and the base station (BTS). Certain factors hinder the maintenance of the CSI accuracy, including Estimation Error, and Aging of the channel [17].

[18] opined that beam-forming is critical to reducing interference for an efficient massive MIMO system. The beam-forming however should have a reduced complexity of implementation. The paper introduced and analyzed a simple, less complex beam-forming model for massive MIMO with respect to high speed train applications.

According to [19], channel state information (CSI) is needed at the transmitter so that the benefits of Massive MIMO can be achieved. Generally, pilot sequences are used to estimate the CSI in wireless networks. The performance of massive MIMO directly depends on the accuracy of channel estimation. In a multi-cell network, when the same or non-orthogonal pilots are used by users in different cells, the base station (BS) will estimate overlapped channel state information (CSI) from these users. Consequently, the precoder based on this estimated CSI will generate interference to these users. This interference is called pilot contamination. To reduce this phenomenon, the pilot sequences of users should be orthogonal within a cell and its neighbors but the number of orthogonal pilot sequences within a specific bandwidth allocation is limited. Based on recent work, as the number of antennas scale up, the following observations need to be considered:

- It becomes most desirable to employ the matched filter as a user detection technique
- The transmission power per user has to be significantly reduced
- Pilot contamination becomes more pronounced and this limits the system performance [20]

It was also hypothetically deduced by [21] that approximately every 10 years, a new mobile communication generation emerges for the improvement of transmission rate, capacity, efficiency and support of more applications. Apart from massive MIMO, certain other technologies deployable for 5G networks were analyzed. This included millimeter wave technology, dense heterogeneous networks, device to device (D2D) communications.

Other related works worth mentioning have also been carried out that focus on energy for sustainable development. Such works include the works of [22–24].

3 Methods

In this section, the research problems (questions) of this study are highlighted, with their corresponding underlying reasons. It also discusses the strategy for searching and filtering out the relevant studies, criteria for the inclusion of any study, assessment of the quality of the papers retrieved and also the degree of relevance of information from each chosen paper.

3.1 Research Questions

This review is aimed at addressing the following research questions and their underlying reasons. This is in order to provide understanding into the deployment of massive MIMO for contribution to the realization of next generation (5G) systems (Table 1 and Fig. 1).

Table 1: Research questions and their corresponding rationale being addressed in the review

Research questions	Rationale
What are the laid-down expectations of 5G wireless networks?	To identify the key specifications of future (5G) networks?
How will Massive MIMO contribute to the realization of the 5G specifications?	To gain knowledge of the operation of massive MIMO in order to understand the roles it will play in realizing the 5G specifications
What are the limitations or challenges associated with 5G?	To identify these limitations which signify improvements in massive MIMO technology for its effective deployment in 5G networks

3.2 Search Strategy

Some keywords related to the subject matter and research problems were used in the search processes. These keywords included: "(MIMO OR massive MIMO OR very large MIMO) AND (5G OR next generation or future generation systems)". Searches in scholarly journals and conferences including Science Direct, IEEE Explore and Google Scholar were carried out in order to obtain the primary studies relevant to the subject matter. Papers considered were from the year 2010 so as to retrieve more current information. Initially, a total of about 500 papers were obtained. After brief checks on the titles, abstracts and conclusions, about 70 of them were filtered. They were then scrutinized to determine those that would be included based on the research focus.

3.3 Inclusion Criteria

After detailed study of these papers, the ones pertaining strictly to irrelevant areas such as pilot contamination and others not related to future (5G) generation systems were removed. Out of the 70 filtered papers, 15 primary studies were selected. These studies are further examined in the next sub-section.

3.4 Assessment of Quality

Each of the fifteen primary studies was thoroughly examined based on three main questions of quality assessment (QA):

- QA1: Does the paper strictly adhere to the essential expectations of a research article in terms of quality of format and presentation, as well as journal specifications?
- QA2: Does it contain sufficient accurate theoretical knowledge of massive MIMO technology especially with respect to 5G networks?
- QA3: Are there mathematical representations or simulations that demonstrate the viability of massive MIMO as a future generation technology?

For each of the primary papers, the scoring metric used in each of the aforementioned questions was the following:

Yes (Y) = 1; Partial (P) = 0.5; No (N) = 0;

4 Results

The selected papers were individually scrutinized and data collected to provide sufficient answers to the already stated research questions. This section shows the results in detail. Table 2 below shows these studies, their source journals and publication years. It is seen that the years range from 2010 through 2017. The results of the assessment for quality for each paper selected are also displayed in Table 3. It is seen from the table that the least score is 1.5 obtained from the assessment of three (3) papers. The maximum score of 3.0 is gotten from only one (1) of the studies. Majority (7) of the papers attained a score of 2.0 while the remaining (4) attained a score of 2.5.

4.1 First Research Question: What are the Laid-Down Expectations of 5G Wireless Networks?

According to telecommunication standards, there are certain main expectations of the future (5G) networks for more efficient communications. 5G is mainly aimed at provision of the following:

- Increased data rate for numerous users, i.e. high capacity and high speed. It must also enable several connections at the same time;
- Improved spectral efficiency at reduced cost compared to fourth generation (4G). It should also be interoperable with Wireless Fidelity (Wi-Fi) and 4G Long Term Evolution to guarantee high data speed;
- Support the internet of things e.g. several million devices.

Other parameters that 5G is expected to improve include lower battery usage, minimized probability of outage, increased number of enabled devices, minimal cost of infrastructure and very reliable communication. The figure below shows the need for more enhanced generation networks as the demand for data communication scales greatly (Fig. 1).

Table 2: Study References, corresponding sources and publication years

References	Source Journals	Year
Lamba and Singh (2014)	Scientific Online Publishing Transactions on Wireless Communications	2014
Neri et al. (2014)	International Journal of Antennas and Propagation	2014
Wang et al. (2016)	Science China Information Sciences	2016
Zheng et al. (2014)	International Journal of Antennas and Propagation	2014
Larsson et al. (2014)	International Journal of Antennas and Propagation	2014
Vardhan et al. (2016)	Indian Journal of Science and Technology	2016
Chen et al. (2017)	Science China Information Sciences	2017
Mishra and Gaur (2016)	International Journal Of Engineering Sciences & Research Technology	2016
Le et al. (2016)	Journal of Mobile Information Systems	2016
Bojkovic et al. (2016)	Advances in Circuits, Systems, Signal Processing and Telecommunications	2016
Kobayashi et al. (2011)	IEEE Trans. Communications	2011
Kaltenberger et al. (2010)	Proc. of Future Network and Mobile Summit	2010
Marzetta (2010)	IEEE Trans. Wireless Communications	2010
Hoydis et al. (2013)	IEEE J. Sel. Areas Communications	2013
Zhang et al. (2013)	IEEE J. Sel. Areas Communications	2013

4.2 Second Research Question: How will Massive MIMO Contribute to the Realization of the 5G Specifications?

Massive MIMO is a form of Multiple User-MIMO system consisting of very many base station (BS) antennas and users. In Massive MIMO, there are several hundreds or thousands of BS antennas that, at the same time, serve tens or hundreds of users in the same time-frequency domain. Some essential properties of Massive MIMO include:

Operation of Time Division Duplexing (TDD): In the case of Frequency Division Duplexing (FDD), the number of BS antennas (M) affects the overhead associated with channel estimation. However, in TDD, the channel estimation overhead does not depend on M. In Massive MIMO, M assumes a very big value, thus TDD is preferable to FDD. For instance, consider a coherence interval of $T = 200$ symbols (this corresponds to coherence time and bandwidth of 1 ms and 200 kHz respectively). Therefore, for FDD systems, the constraint on the number of BS antennas and users is $M + K < 200$, whereas for TDD systems, such constraint is $2K < 200$, where K represents number of simultaneous active users.

The graph below illustrates the feasible regions (M,K) for both TDD and FDD systems. It shows that the TDD region is much higher than the FDD region. With TDD, increasing the number of antennas is independent of the channel estimation resources (Fig. 2).

Table 3: Evaluation of quality for each selected study

References	QA1	QA2	QA3	Total Score
Lamba and Singh (2014)	Y	P	N	1.5
Neri et al. (2014)	Y	P	Y	2.5
Wang et al. (2016)	Y	Y	N	2.0
Zheng et al. (2014)	Y	Y	Y	3.0
Larsson et al. (2014)	Y	Y	N	2.0
Vardhan et al. (2016)	Y	P	Y	2.5
Chen et al. (2017)	Y	P	P	2.0
Mishra and Gaur (2016)	Y	P	P	2.0
Le et al. (2016)	Y	P	N	1.5
Bojkovic et al. (2016)	Y	P	N	1.5
Kobayashi et al. (2011)	Y	P	P	2.0
Kaltenberger et al. (2010)	Y	P	P	2.0
Marzetta (2010)	Y	P	P	2.0
Hoydis et al. (2013)	Y	P	Y	2.5
Zhang et al. (2013)	Y	P	Y	2.5

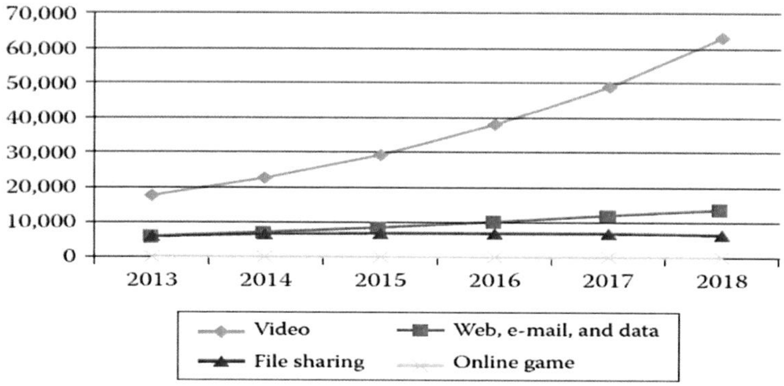

Fig. 1: Yearly demand for transmission over Internet Protocol (IP). Values are in petabit / month.

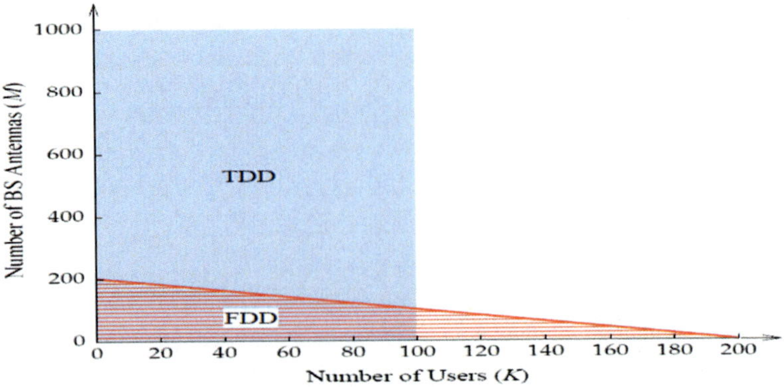

Fig. 2: The feasible regions (M,K) for FDD and TDD systems, for a coherence interval T = 200 symbols.

Linear processing: Due to the many BS antennas (M) and active users (K), the signal processing at both ends is inevitably associated with large dimensional matrices/vectors. This introduces more complexities to the system's operations. Thus in massive MIMO simple linear signal processing (uplink linear combination techniques and downlink linear precoding techniques) is preferable.

Favorable propagation: This implies that the coefficient matrix of the channel between the antenna array of the base station (BS) and the users is well-conditioned i.e. minimal inherent noise. In Massive (very large) MIMO, under some conditions, the property of favorable propagation stands based on the law of large numbers.

Physical size of antenna array of the BS: For instance, in a with 128-antenna cylindrical array which comprises four circles of 16 dual-polarized antennas, the adjacent separation of the antennas is about 6 cm (half a wavelength) at a frequency of 2.6 GHz and thus, this array has only a physical size of 28cm × 29cm. That is, the BS array need not have a large physical size.

Scalability: In Massive MIMO, under time division duplexing (TDD) operation, the BS learns the channels via uplink training. The time needed for channel estimation does not rely on the number of BS antennas. Hence, regardless of how numerous the BS antennas can be, it has no effect on the channel estimation overhead. Also, very simple signal processing takes place at each user and is independent on users' activity or inactivity.

4.2.1 Operation of Massive MIMO

TDD operation is a preferable option in Massive MIMO. Three operations take place in a coherence interval: uplink data transmission, downlink data transmission and channel estimation (including the training process in both uplink and downlink). The figure below shows a Time Division Duplexing (TDD) Massive MIMO protocol (Fig. 3).

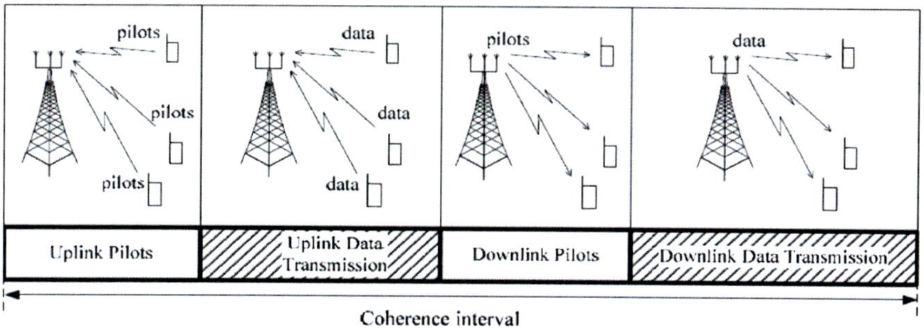

Fig. 3: TDD Massive MIMO Transmission protocol

4.2.2 Channel Estimation

The Channel State Information (CSI) obtained through the uplink training is needed by the base station (BS) for uplink detection of the user signals and downlink precoding of these signals. Every user has an orthogonal pilot sequence which is sent to the BS. The BS then performs channel estimation using these sequences. Also, each user via downlink training acquires and uses partial knowledge of the CSI to detect transmitted BS signals. The BS beamforms signals to active users during a short time using linear precoding schemes.

4.2.3 Uplink Data Transmission

A portion of the coherence interval is applied in the uplink data transmission. During uplink, signals are transmitted from all K active users to the BS in the same frequency-time resource. Then the BS detects these transmitted signals using the channel estimates and linear combining methods.

4.2.4 Downlink Data Transmission

During downlink, signals are sent from the BS to all K active users in the same frequency-time resource. More specifically, the BS uses its channel estimates combined with user-intended symbols to create M precoded signals which are then sent to M antennas.

4.3 Third Research Question: What are the Limitations or Challenges Associated with 5G?

A: Reciprocity of Channel: Time Division Duplexing (TDD) operation is dependent on the reciprocity property of the channel. It is generally assumed that the propagation channel is basically reciprocal provided the propagation is not distorted by interfering magnetic fields. The hardware circuitry in the transceivers of terminals and base station may however not be reciprocal between the downlink and uplink. Hardware-circuitry calibration does not really pose any challenge and there are already-tested calibration-based results in practice [3, 12].

B: Pilot Contamination: In a Massive MIMO system, every terminal has assigned to it an orthogonal uplink pilot sequence. However the number of orthogonal pilot sequences is constrained by the ratio of the coherence interval duration and delay-spread of the channel. For a typical case as in [13], the there is a peak of about 200 orthogonal pilot sequences for a coherence interval of 1ms. The supply of these pilot sequences that is available in a multi-cellular system is very exhaustible. The associated negative consequence of pilot re-use in different cells is termed "pilot contamination". Pilot contamination is really not unique to massive MIMO, but is considered by many researchers as a major challenge to the system especially when the number of antennas is proliferated, at least with receivers that depend on channel estimation that is pilot-based [13, 14].

C: Unfavourable Propagation: Massive MIMO functions perfectly under ideal propagation environments. But, some propagation environments may exist in practice with adverse properties. A good example is a propagation environment where the channels from the BS to the different users share some common scattering properties. Such channel is regarded unfavorable. One possible solution to this challenge is large-area distribution of the BS antenna arrays.

D: Need for New Designs and Standards: Deploying Massive MIMO in practical systems like LTE will make it more efficient. However, the major limitation of LTE standard is that only a maximum of 8 antenna ports are allowed at the base station [2]. In addition, LTE makes use of assumed channel information. For instance, in the LTE downlink, the BS transmits the control signals to the users via multiple fixed beams. Then feedback from the users is sent to the BS via the most efficient beam. Such beam will then be selected and used by the BS for downlink transmission. On the contrary, Massive MIMO uses measured (estimated) channel information. Therefore, new standards are needed for practical implementation of Massive MIMO. Related hardware designs should also be considered.

4.4 Ongoing research problems regarding Massive MIMO technology

While massive MIMO provides solutions to numerous conventional communication theory problems, it reveals a number of unexplored problems that need the attention of further studies:

A. Distributed Massive MIMO

Analyses on centralized antenna arrays have been carried out by researches and their shortcomings revealed. Current researches have also explored the benefits of a distributed radio system. One such benefit is improved spatial diversity. Also, distributed Massive MIMO supports easier placement of antenna elements e.g. distributing them among several buildings. However, fitting hundreds of antennas in one array especially at low frequencies is laborious and detrimental to the efficiency of the system.

B. High Mobility

Mobility issues are more critical in Massive MIMO systems than traditional wireless systems. Massive MIMO depends on current CSI to beamform signals to every user and the accuracy of the beam forming may be prevented by shorter channel coherence time of a mobile user. The coherence time of the channel depends on the mobile user's speed, and the higher the speed the more the rate of CSI update. This challenge is encountered more particularly in time division duplex operations. More research needs to be undertaken on the modifications of the physical layer of Massive MIMO to make it deployable in high-mobility systems like vehicular-based UEs.

C. Antenna Patterns

So far, works on Massive MIMO have been carried out with omnidirectional or patch antennas in either linear or 2-dimensional planar array with uniform spacing. However, the antenna patterns that best suit the performance of Massive MIMO still remains inconclusive. Further work needs to be conducted using distributed and curved (cylindrical and spherical), or more extensive use of directional antennas to arrive at the best option for optimum performance.

D. Hybrid Beam Forming

It is more energy-intensive and expensive to have separate transceiver circuitry chains for each antenna in a multi-user MIMO system. Hybrid beam forming reaps the benefits of digital beam forming and also improves power consumption and cost. Further research in this area will be targeted on developing the algorithms, performance characteristics, and energy optimization for a hybrid beam forming approach and results compared with those of existing methods.

E. Energy Efficiency

Since hundreds of transceivers are located in the BS, limiting the radiated power becomes a serious issue. This is especially so with digital-to-analog converters and analog-to-digital converters of high speed. New algorithms need to be developed to reduce energy by reducing sample rates, selectively impairing antennas or using other power reduction strategies.

F. Improvements in Channel State Information Feedback

It is very necessary for base stations in Massive MIMO to frequently receive CSI from the user equipment (UE). However the CSI feedback transfer takes a significant part of the time needed for data transmission. Furthermore no beam forming can occur until the CSI feedback is received by the base station, and this can be a threat to the operational efficiency of the system. New methods, such as CSI quantization strategies or beam forming CSI reference signals to the base station are potential solutions to these problems.

G. FDD Massive MIMO

Most researches on Massive MIMO are with respect to time division duplexing (TDD) operations. TDD is preferable because it reduces CSI overhead by exploiting reciprocity of the channel. But with frequency division duplex (FDD), as the base station antennas proliferate the CSI overhead also increases. However, there are several benefits associated with FDD operation, which make it useful in existing wireless networks. Efficient adaptation of Massive MIMO with FDD for optimum performance would be a very critical area of research.

5 Conclusion

Owing to its huge potentials especially with respect to spectral efficiency, energy efficiency, reliability and robustness, massive multiple-input multiple-output (MIMO) is regarded in many previous and on-going researches as a principal enabling technology for the emerging cellular systems beyond fourth generation (4G). It is a system, with numerous antenna elements installed at the base station which guarantees enhanced channel reliability with very good coverage. The dramatic improvement in the energy and spectral efficiency of wireless systems is largely due to the bulky antenna arrays at the base station. It allows for the massive deployment of parallel low-power, low-cost hardware both at the transmitter and receiver side, in place of large, expensive and power-intensive hardware. Challenges still exist that pertain to massive MIMO, including the key examples highlighted in this paper. As research continues to unfold, more solutions and models are being proposed to effectively tackle these challenges for the full realization of the potential power of massive MIMO for future 5G cellular technology.

Acknowledgement. The authors appreciate the sponsorship from Covenant University through its Centre for Research, Innovation and Discovery, Covenant University, Ota Nigeria.

References

1. Kachhavay, M.G., Thakare, A.P.: 5G technology-evolution and revolution. Int. J. Comput. Sci. Mob. Comput. **3**, 1080–1087 (2014)
2. Lu, L.: An overview of massive MIMO: benefits and challenges. IEEE J. Sel. Topics Sig. Process.**8**(5) 742–758 (2014)
3. Prinima, D., Pruthi, D.J.: Evolution of Mobile Communication Network: from 1G to 5G. Int. J. Innov. Res. Comput. Commun. Eng.**4**, 224–227 (2016)
4. Meraj, M., Kumar, S.: Evolution of mobile wireless technology from 0G to 5G. Int. J. Comput. Sci. Inf. Technol. **6**(3), 2545–2551 (2015)
5. Vardhan, P., Gupta, M., Kumar, A.: Massive-MIMO- past, present and future: a review. Indian J. Sci. Technol. **9**(48), 1–13 (2016)
6. Larsson, E.G., Edfors, O., Tufvesson, F., Marzetta, T.L.: Massive Mimo For Next Generation Wireless Systems, arXiv:1304.6690v3 [cs.IT], pp. 1–20 (2014)
7. Agiwal, M., Roy, A., Saxena, N.: Next generation 5G wireless networks: a comprehensive survey. IEEE Commun. Surv. Tutor. **18**(3), 1617–1655 (2016)
8. Le, N.T., Hossain, M.A., Islam, A., Kim, D., Choi, Y., Jang, Y.M.: Survey of promising technologies for 5G networks. Mob. Inf. Syst. **2016**, 1–25 (2016) Article ID 2676589
9. Wang, C.-X., Shangbin, W., Bai, L., You, X., Wang, J., Chih-Lin, I.: Recent advances and future challenges for massive MIMO channel measurements and models. Sci. Chin. Inf. Sci. **59**(2), 1–16 (2016)
10. Lamba, M., Singh, C.: Challenges and Future Direction for MIMO Communication Systems. Sci. Online Publishing Trans. Wirel. Commun. **1**(2), 42–50 (2014)
11. Zhang, J., Yuan, X., Ping, L.: Hermitian precoding for distributed MIMO systems with individual channel state information. IEEE J. Sel. Areas Commun. **31**(2), 241–250 (2013)
12. Neri, M., et al.: Ultra-broadband mobile networks from LTE-Advanced to 5G: evaluation of massive MIMO and multi-carrier aggregation effectiveness (2017)

13. Zheng, K., Ou, S., Yin, X.: Massive MIMO channel models: a survey. Int. J. Antennas Propag. **2014**, 1–10 (2014) Article ID 848071
14. Liu, W., Han, S., Yang, C., Sun, C.: Massive MIMO or small cell network: who is more energy efficient? In: Wireless Communications and Networking Conference Workshops (WCNCW), 2013, pp. 24–29. IEEE (2013)
15. Chen, X., Lu, J., Fan, P.: Massive MIMO Beam-forming for High Speed Train Communication: Directivity vs Beam width", arXiv:1702.02121v1 [cs.IT], (February 2017)
16. Mishra, A.K., Gaur, S.: Review of the pilot contamination problem for massive MIMO and possible solution. Int. J. Eng. Sci. Res. Technol. **5**(7), 538–541 (2016)
17. Chin, W.H., Fan, Z., Haines, R.: Emerging technologies and research challenges for 5G wireless networks. IEEE Wirel. Commun. **21**(2), 106–112 (2014)
18. Bojkovic, Z., Bakmaz, B., Bakmaz, M.: Recent trends in emerging technologies toward 5G networks. Adv. Circ. Syst. Sig. Process. Telecommun. 137–143 (2015)
19. Kobayashi, M., Jindal, N., Caire, G.: Training and feedback optimization for multiuser MIMO downlink. IEEE Trans. Commun. **59**(8), 2228–2240 (2011)
20. Kaltenberger, F., Haiyong, J., Guillaud, M., Knopp, R.: Relative channel reciprocity calibration in MIMO/TDD systems. In: Proceedings of Future Network and Mobile Summit (2010)
21. Marzetta, T.L.: Noncooperative cellular wireless with unlimited numbers of base stationantennas. IEEE Trans. Wirel. Commun. **9**(11), 3590–3600 (2010)
22. Adeyemi-Kayode, T.M., Misra, S., Damaševičius, R.: Impact analysis of renewable energy based generation in west Africa – a case study of Nigeria. Problemy Ekorozwoju **16**(1), 67–78 (2021)
23. Adesanya, A., Misra, S., Maskeliunas, R., Damasevicius, R.: Prospects of ocean-based renewable energy for West Africa's sustainable energy future. Smart Sustain. Built Environ. **10**(1), 37–50 (2021)
24. Ayodele, E., Misra, S., Damasevicius, R., Maskeliunas, R.: Hybrid microgrid for microfinance institutions in rural areas–a field demonstration in West Africa. Sustainable Energy Technol. Assess. **35**, 89–97 (2019)

Deep Networks for Brachial Plexus Nerves Segmentation and Detection Using Ultrasound Images

Kuldeep Pisda, Prince Jain, and Dilip Singh Sisodia(✉)

National Institute of Technology Raipur, Raipur, India
dssisodia.cs@nitrr.ac.in

Abstract. The Brachial Plexus (BP) nerves extend from the spinal cord, through the cervicoaxillary canal in the neck, over the first rib, and armpit. BP nerves are mainly responsible for the efficient working of our upper limbs. During any neck surgery, catheter placement is very crucial in precisely locating the BP nerves. The catheter's slight misplacement may permanently damage BP nerves resulting in Erb's Paralysis and muscle paralysis or may take a too long time to regain full patient strength. An automated assistive technique for locating the BP nerves may help medical practitioners detect BP Nerves faster before surgery. In this article, deep neural networks such as a fully convolutional network (FCN), ResNet-34, MobileNet, DenseNet, and LinkNet are empirically evaluated to segment and locate BP nerves from given ultrasound images. To reduce the training time and computation cost, transfer learning is used to train the networks. Experiments are performed using the Kaggle dataset of BP nerves ultrasound images. The performances of networks are measured using accuracy, sensitivity, and specificity. Intersection over Union (IoU) score or Jaccard index is also computed for performance analysis. Results show that LinkNet is the fast and most efficient transfer learning method among the used networks and ultrasound images.

Keywords: Brachial Plexus nerves · Image segmentation · Deep learning · DenseNet · MobileNet · ResNet-34 · LinkNet

1 Introduction

Misplacement of a catheter causes severe pain and is not ideal for the patients. Narcotics and anesthesia help in reducing the pain. In local anesthesia, an anesthetic is used in part of a body to make it unconscious. However, it is difficult to identify target regions to operate as it varies with the person. Medical imaging is the most widely known method for diagnostics & treatment purposes. There are various ways to go about it, including X-ray, magnetic resonance imaging (MRI), and computed tomography (CT) scans. However, the one that stands out is ultrasound sonography (USG) [1]. Ultrasound imaging has its advantages in medical imaging. It is cheaper than other methods; it has a low maintenance cost and safer than its other alternatives, like X-ray, CT scan. The USG

L. Garg et al. (Eds.): ISMS 2020, LNNS 303, pp. 132–146, 2022.
https://doi.org/10.1007/978-3-030-86223-7_13

devices use high-frequency sound waves, which provide high resolution and incredible detail in musculoskeletal problems compared to others like MRI.

However, the quality of a produced image in ultrasound is low and requires appropriate further processing before diagnosis. Regional anesthesia helps in operating with nerves [2]. Previously, it was highly dependent on the needle blindly to pinpoint the target block. However, with the advancements, a new technique called ultrasound-guided regional anesthesia is used. It provides direct visualization, and operation is no longer blind. It has tremendously reduced the pain experienced by the patient during and post-surgery. It takes hours of practice to get expertise in this technique, and there are not many practitioners with such vast knowledge.

Even today, the Brachial Plexus nerve localization is done manually by practitioners and is often prone to human error as it depends on the surgeon's skill. BP Nerves are specific visual patterns in the given ultrasound image; hence we may use artificial neural networks-based techniques to segment BP nerves in visible patterns from the given ultrasound image [3]. This paper aims to assist medical practitioners in high accuracy localization of Brachial Plexus nerves. These nerves begin in the neck and spread in the entire upper body. We may use computer vision algorithms to detect BP Nerves in less time with fewer human errors. It will reduce human errors in detecting the nerves, resulting in the catheter's correct placement, which eases the patient's experience in resulting pain [4, 5].

With recent advancements in data and image processing techniques, deep learning is now solving many real-world problems in various domains. Deep learning convolutional neural network (CNN) architectures have tremendous potential in pattern recognition and medical image analysis. This paper relies on the CNN-based approach using U-net architecture and light-weight state of the art LinkNet architecture for medical imaging analysis [6–8]. It will perform classification and segmentation/localization of nerves on the provided dataset to make better predictions.

Deep learning methods require considerable data, time, and computing power to give good results. Therefore, we used transfer learning instead of making the model from scratch. It outperforms the same model trained from scratch [9, 10].

2 Literature Review

Olaf, Ronnenberger et al. [5] introduced U-Net, a convolutional network for image segmentation, especially in the biomedical field. It relies on two paths, expansion, and contraction, making localization of desired objects and accurately identifying individual pixels. The successive expansion and contraction levels, using the upsampling operator, ensured that the training worked even with a small number of training images. The transfer of information in the higher resolution layers in the unsampling made all the difference. It relies on stochastic gradient descent. Augmentation used on a large scale with decent training time meant that the results were reliable. The U-Net is still the base for segmentation in Medical Images.

Ahmed et al. [11] performed segmentation on breast ultrasound images to accurately identify and treat diseases associated with it using thresholding. The paper proposed a scale-invariant feature transform (SIFT) method. The method used descriptors for training on neural networks to identify critical points and regions in an image.

Zahra Shobhaninia et al. [8] used segmentation to measure and find fetal head circumference from Ultrasound images during pregnancy. However, missing boundaries, blur, and noise are challenges associated with such ideas. The paper suggests using U-net made use of a deep convolutional neural network called LinkNet for performing multitask and shows promising results. It performs segmentation tasks and uses the Ellipse turner module on 999 Ultrasound images, and each is associated with different loss functions.

Hadjerci et al. [12] use active contour to perform segmentation in Ultrasound images. This comparative study uses different parts of the body and compares various factors like efficiency and reliability of every framework. The first phase uses the Hierarchical multi-affine algorithm (HMA) for de-speckling images. The second step is feature extraction to remove data of nerves with ambiguities. The classification was performed to separate nerve tissues from the ones where it is not present. Comparison of various architectures, filter methods, feature selection, and preprocessing was also done. The last stage was segmentation to assist ultrasound-guided regional anesthesia (UGRA).

Vidushi et al. [13] highlight difficulties in nerve segmentation of ultrasound images and improve the existing solution. Three approaches for image processing and filters to reduce noise were used. After processing, images were divided into training and test images, converted to 2D arrays. After image segmentation, a support vector machine (SVM) predicts new images' nerve areas. The uncanny operators were used, but no use of transfer learning [5].

Akhilesh Kakade et al. [2] used supervised learning methods to perform semantic nerve segmentation on Ultrasound images to identify and detect Brachial Plexus nerves. To determine the region of interest, deal with noise, and see edges Gabor filter was used. In this work, pixel classification was done using a linear binary pattern and sliding window techniques. It takes input from 0 to 255 and performs binary classification. The best dice score obtained 0.68 using U-Net architecture with post-processing.

3 Dataset

The used dataset [14] has 11272 images with ultrasound images, X-labels, Y-labels, and binary masked images. The X-label are grayscale images of dimensions 580X420 with values ranging from 0–255 & binary mask images of the same size containing only 2-pixel values. These images are of 47 different subjects. The annotations of nerve segments in data were marked manually and subjected to error, meaning two very similar ultrasound images contradicting Brachial Plexus nerves' presence. Figure 1 shows the different types of images & their annotations. The testing set contained a total of 1000 images and the confusion matrix is made on the testing set.

4 Proposed Methodology

The proposed methodology is based on supervised learning algorithms. Here, the region of interest is BP nerves. The challenge is the accurate segmentation from ultrasound images containing other artifacts like bones, vessels [15, 16]. The even bigger issue is the presence of noise present in ultrasound images, leading to errors in the training process [17]. We used SVM algorithm, to show cast inefficiency to solve the segmentation problem [18, 19]. We also used transfer learning by employing architectures like fully convolutional network (FCN), ResNet-34, MobileNet, DenseNet, and LinkNet to test and analyze various networks dataset. In the end, we have performed an empirical analysis. The proposed methodology is carried out in five significant steps, as sown in Fig. 2. A detailed description of each step is given in subsequent sections.

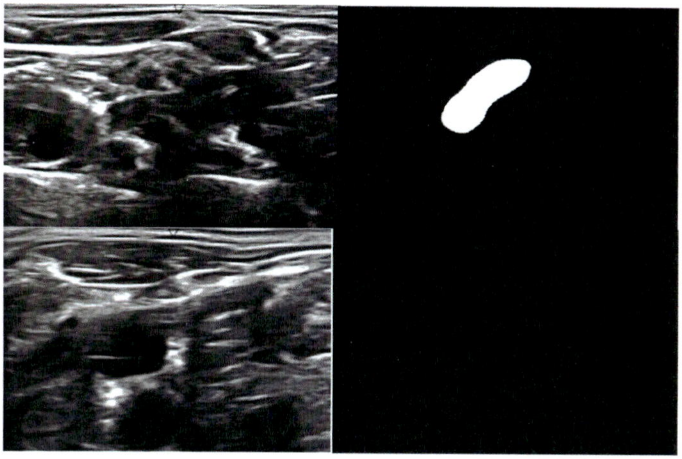

Fig. 1. Dataset containing the segmented image

4.1 Anomaly Removal

As shown in Fig. 3, the data set images consist of multiple anomalies, such as two similar ultrasound images having different masks. One is establishing the presence of the Brachial Plexus nerve, and another one not. Experts marked annotations, and any marked human data is error-prone and noises. We removed the contradictory image with the absence of a mask. It was done by dividing the image into blocks of 20 × 20 and calculating each block's intensity's histogram. The similarity is the cosine distance between the vectorized intensity of blocks of two pictures, as stated in Eq. (1).

$$cos\theta = \frac{\vec{a} \cdot \vec{b}}{|\vec{a}||\vec{b}|} = \frac{\sum_1^n a_i b_i}{\sqrt{\sum_1^n a_i^2}\sqrt{\sum_1^n b_i^2}} \tag{1}$$

Where $\vec{a} \cdot \vec{b} = \sum_1^n a_1 b_i = a_1 b_1 + a_2 b_2 + \cdots + a_n b_n$ is the dot product of the two vectors. The similarity threshold was set at 80%, meaning two images with 80% or more similarity and different masks will lead to the anomaly, and we removed one of the pictures. We repeated this step for every patient. This threshold reduced total images to 3906. This step was essential for improving the accuracy and efficiency of any algorithm in training.

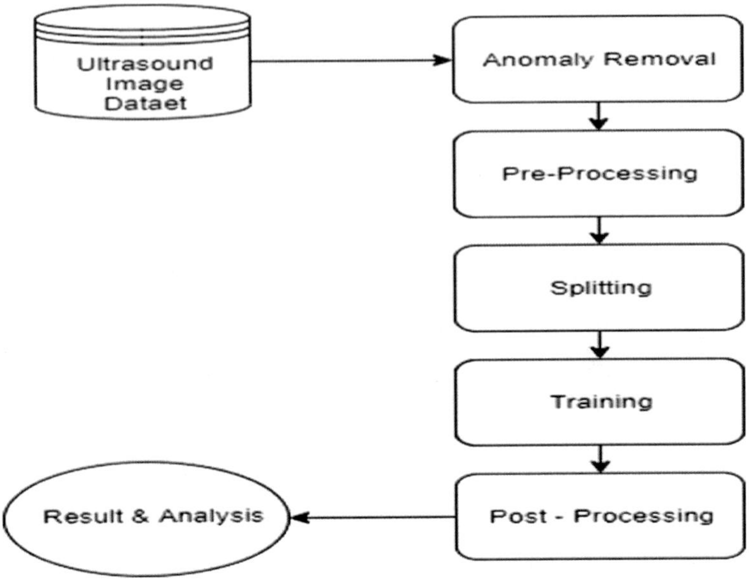

Fig. 2. The workflow of the analysis

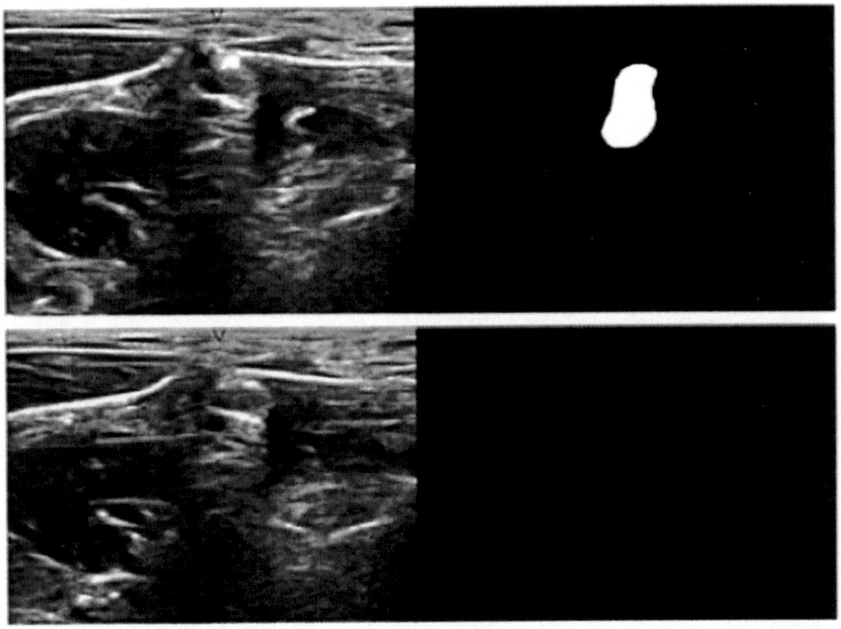

Fig. 3. Anomaly in the dataset

4.2 Preprocessing

First, all the used images were resized to a dimension of 224 × 224 since most of the networks, i.e., FCN, MobileNet, and DenseNet, taken images as input with the same dimension [15, 20]. We used augmentation to prevent the machine from over-fitting. Ultrasound images come with a lot of noise, which was denoised using Gaussian Blur. It removes noise while keeping edges sharp [4, 15]. After this, we performed histogram equalization to improve the image's contrast and spread pixel distribution for better training [21]. Figure 4 depicts the preprocessing workflow.

4.3 Training and Testing

We divided the images dataset into the ratio of 90:10 in the training set & testing set. It is done for better training and evaluating the model in each phase [21].

First, we performed a simple binary classification of image dataset using support vector machine (SVM) classifier. Figure 5 shows the analysis and flow diagram of the same, and Fig. 6 shows the results. [22].

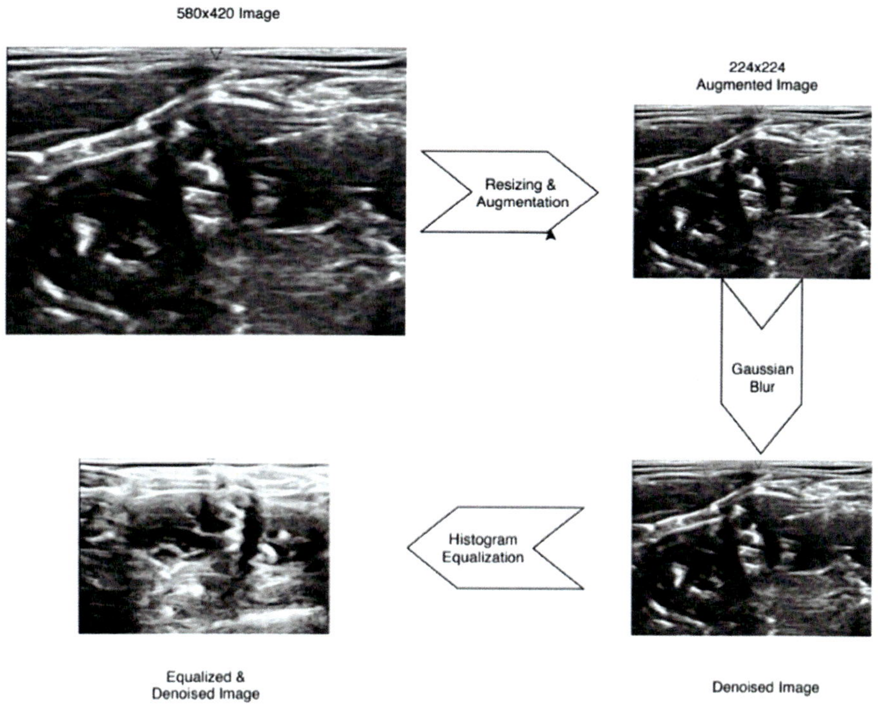

Fig. 4. Pre-processing of the dataset

Fully Convolutional Network (FCN): Figure 7 shows the basic working of FCN, which is structurally the same as CNN only with one significant difference. In FCN,

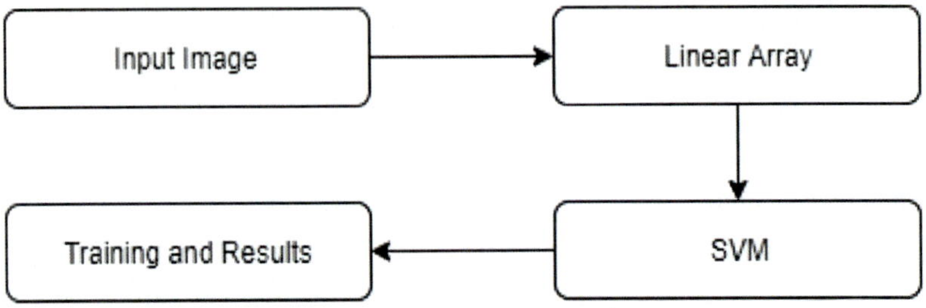

Fig. 5. Workflow for SVM

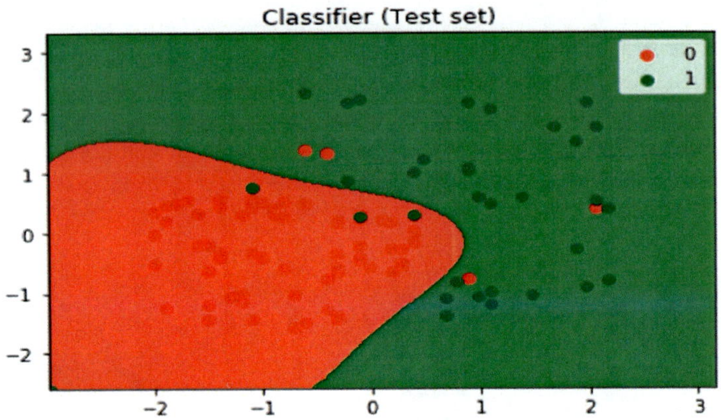

Fig. 6. Semantic segmentation using SVM classifier

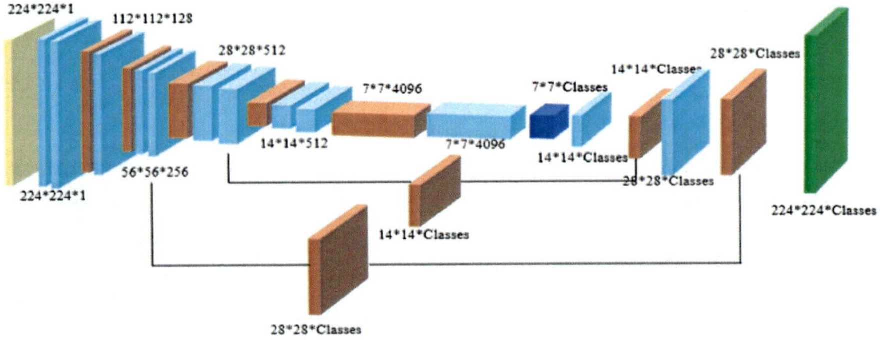

Fig. 7. FCN Architecture

we substitute the last fully connected layer with another convolutional layer with a large receptive field [23, 24]. This convolutional layer has a larger perception, which helps in semantic segmentation, i.e., identifying every pixel.

ResNet-34: ResNet architecture is also called a Residual Network, very popular in the field of Image Segmentation. It has lots of versions available like Resnet-19, ResNet-34, ResNet-50, and ResNet-101 [24, 25] etc.

We used transfer learning because these architectures are robust. These architectures have parameters in 10s of millions.

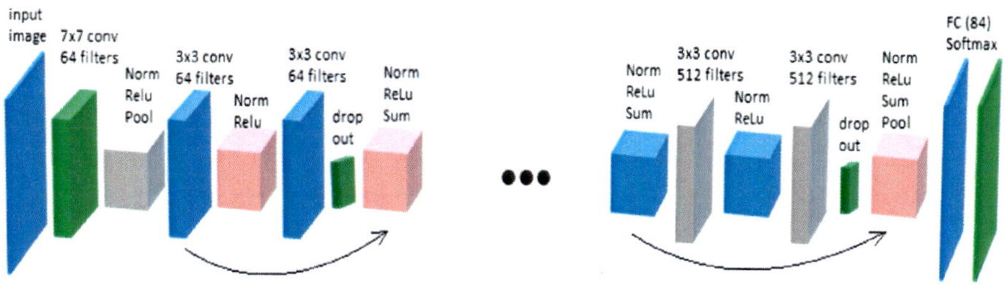

Fig. 8. ResNet architecture [25]

The difference between typical CNN and ResNet is that in ResNet, we add the Previous block's output with the current block's output for RELU operation. It is a bit more complicated but helps extract features better [26]. Figure 8 shows the architecture of ResNet.

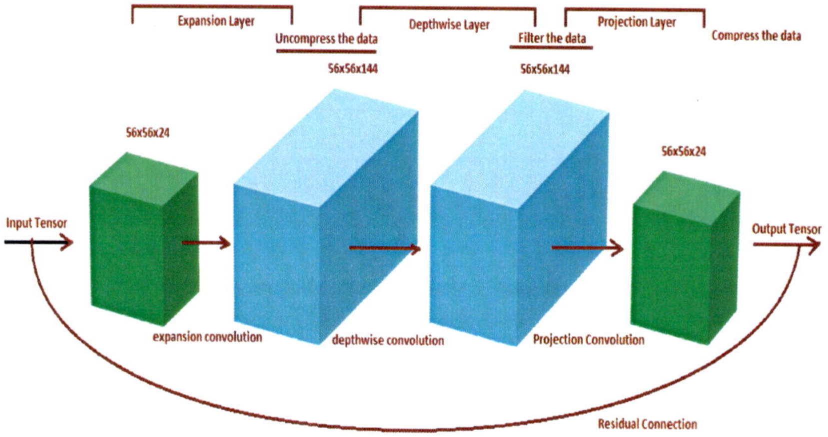

Fig. 9. MobileNet architecture [27]

MobileNet: Andrew G proposed this architecture (Fig. 9) in April 2017 [27]. Compared to previous architectures, it is light weight. It is also a Deep Network with lots of layers of architecture. Embedded devices with limitation of memory primarily use MobileNet.

LinkNet: Various embedded devices use this algorithm. It can perform real-time semantic segmentation. We use it mostly in training self-driving cars [26]. It has a vast implementation in Virtual Reality may also be implemented with the help of Transfer Learning.

The LinkNet has lots of encoding and decoding stages connected, as shown in Fig. 10 above. It disintegrates the image. Later it is passed to 3 × 3 Conv2D layers. The structure of the network reduces the number of parameters [28]. We have used ResNet34 encoder-decoder for our problem.

We use Batch Normalization in each Convolutional Layer. Its output goes directly to Rectified Linear Unit. The left half of the architecture shown in Fig. 10 is the encoder, while the right is the decoder.

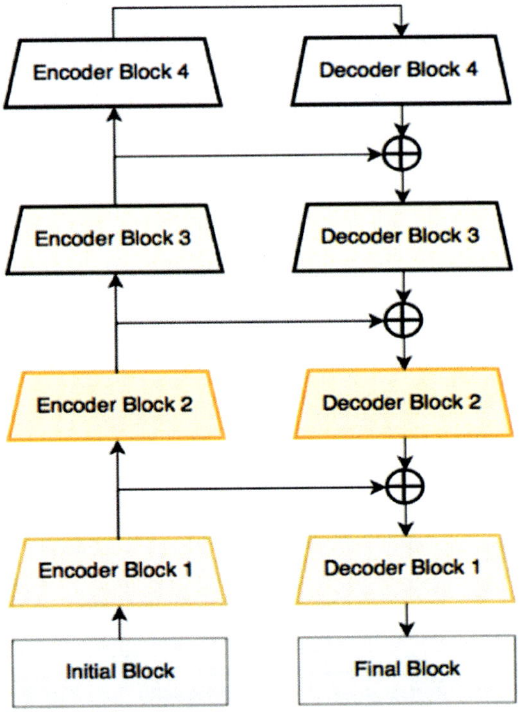

Fig. 10. Encoder-decoder blocks of LinkNet [28]

DenseNet: Cornell Uni and Facebook Research in late 2017 proposed this architecture [29]. It has several versions such as, DenseNet-121, DenseNet-169, DenseNet-201 and DenseNet-264. We used DenseNet-121. It uses the concept of residue like in residual architectures but in an optimal way. Instead of passing all the previous block inputs, only a few are passed but still forwarding all the relevant information. We may implement this architecture (Fig. 11) with the help of transfer learning.

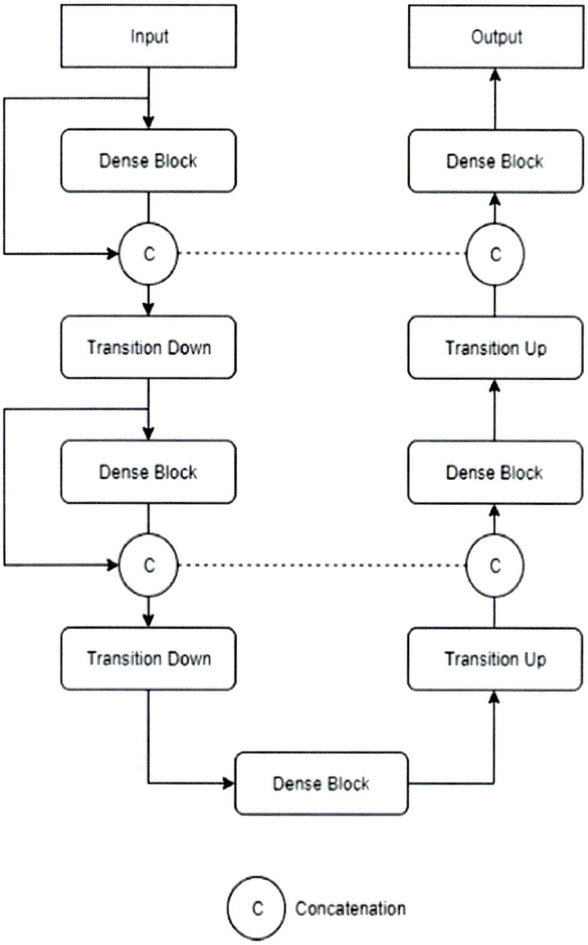

Fig. 11. DenseNet workflow [29]

5 Results

Experiments are performed after removing the anomalies and inconsistent images in the original dataset, the dataset consider for experimentation is shown in Table 1 which consists of total of 3906 images.

Table 1. Distribution of BP and Non-BP nerves in dataset after removing the anomalies

Type of images	Number	Percentage
Normal images (without BP Nerves)	1467	37.5%
Images with BP Nerves	2439	62.4%

This size of dataset may not be sufficient for better training of deep learning networks, therefore we performed data augmentation, and increased the dataset size to 9970 images. The 9970-image dataset is divided into a ratio of 90:10 ratio of the training set to the testing set. The testing set contained around total of 997 images and the used models' performance is evaluated on the testing set. We performed the analysis of the ultrasound images on six architectures. We used a batch size of 4 with ten iterations with a callback, and each iteration containing 20 steps for training. It took varying times for each architecture.

5.1 Confusion Matrix

The confusion matrix is a square matrix containing the prediction result of a classification problem. Various fields show correct and wrong values in numbers. The confusion matrix is used to derive various classification performance measures. The problem is not only to identify BP nerves but also to segment them. However, we need to segment the BP nerves first. We classified the images based on the number of nerves they are segmenting. We used six different learning algorithms to solve the given problem (Fig. 12).

SVM	Predicted True	Predicted False
Actual True	487	135
Actual False	87	288

FCN	Predicted True	Predicted False
Actual True	523	99
Actual False	72	303

ResNet34	Predicted True	Predicted False
Actual True	492	130
Actual False	74	301

MobileNet	Predicted True	Predicted False
Actual True	558	64
Actual False	58	327

DenseNet	Predicted True	Predicted False
Actual True	466	156
Actual False	38	337

LinkNet	Predicted True	Predicted False
Actual True	566	56
Actual False	47	328

Fig. 12. Confusion matrix of various algorithms

5.2 Specificity, Sensitivity and Accuracy

The confusion matrix tells us about the division of actual and predicted numbers. These numbers only are not enough to know about a model's efficiency. Therefore, we used other derived performance measures such as accuracy, sensitivity, and specificity to measure the efficacy of used architectures elegantly.

Sensitivity: Sensitivity [26] measures the sample which the model has correctly identified as positives from total positives. It can be calculated using Eq. (2).

$$Sensitivity = \frac{TP}{P} \tag{2}$$

TP indicates the number of true positive results (Number of samples identified correctly as positives), and P represents the total number of positive samples in the set.

Specificity: Specificity [26] is a ratio of predicted negative to total negative. It can be calculated using Eq. (3).

$$Specificity = \frac{TN}{N} \tag{3}$$

TN indicates the number of true negatives results (Number of samples identified correctly as negatives), and N represents the total number of negative samples in the set.

Accuracy: It tells how accurate a system is. It is the ratio of total corrected predictions with all predicted values. It can be calculated using Eq. (4).

$$Accuracy = \frac{TP + TN}{P + N} \tag{4}$$

Table 2 shows the performance summary of all used models on these three measures using BP nerve datasets.

Table 2. Summary of performance measures for used models

Models	Sensitivity	Specificity	Accuracy
SVM	78.29	76.80	77.73
FCN	84.08	80.80	82.84
ResNet-34	79.09	80.26	79.53
MobileNet	89.71	87.20	88.76
DenseNet	74.91	89.86	80.54
LinkNet	90.99	87.46	89.66

5.3 Intersection Over Union (IoU) Score or Jaccard Index

Deep architectures models' performance is also measured using Jaccard loss (also known as intersection over union (IoU) loss. The IoU score is a ratio of intersection of true pixels and predicted pixels with the union of both. Any system associated with a prediction bounding region or box can incorporate IOU for evaluation purposes. Dividing the area/region of cross-overlap to the area/region of union provides us with an IoU score. Table 3 shows a summary of training, testing, validation losses, and IoU scores of different used deep architecture models.

Table 3. Summary of losses and IoU scores for used deep architecture models

Architectures/ Deep networks	Training loss	Training IoU	Validation Loss	Validation IoU	Testing Loss	Testing IoU
FCN	0.53	0.45	0.58	0.43	0.55	0.48
MobileNet	0.42	0.59	0.59	0.44	0.51	0.51
DenseNet	0.45	0.56	0.56	0.46	0.54	0.49
LinkNet	0.46	0.55	0.58	0.45	0.50	0.52
ResNet-34	0.57	0.48	0.59	0.43	0.60	0.42

6 Conclusion

In this article, different deep architectures were empirically for the segmentation and detection of BP Nerves using an ultrasound image dataset. Rather than designing and training the deep architectures from scratch, transfer learning saves time and computation cost. First, ultrasound image preprocessing was performed, and semantic segmentation was done using SVM classification. The batch size of 4 was used with ten epochs. The algorithms were also tested without preprocessing, but we lost some images' visual properties while applying blur and smoothening. Five different deep neural networks, such as a fully convolutional network (FCN), ResNet-34, MobileNet, DenseNet, and LinkNet, were evaluated. The used deep networks performance was measured using accuracy, sensitivity, and specificity. The intersection over union (IoU) score was also calculated to measure the loss. The LinkNet model with encoder and decoder outperforms all other used models.

In this work, experiments were performed using Keras and TensorFlow as a back-end. In the future, it can be tested using other deep learning libraries such as Pytorch, known to perform better against TensorFlow with the same ultrasound images dataset.

References

1. Chan, V.W.S., Perlas, A., Rawson, R., Odukoya, O.: Ultrasound-guided supraclavicular brachial plexus block. Anesth. Analg. **97**(5), 1514–1517 (2003). https://doi.org/10.1213/01.ANE.0000062519.61520.14

2. About Brachial Plexus and Traumatic Nerve Injury, Hospital for Special Surgery (January 2020). https://www.hss.edu/brachialplexuscenter-about.asp
3. Singh Sisodia, D., Verma, S.: Image pixel intensity and artificial neural network based method for pattern recognition. World Acad. Sci. Technol. **57**, 742–745 (2011)
4. Thatte, M.R., Babhulkar, S., Hiremath, A.: Brachial plexus injury in adults: diagnosis and surgical treatment strategies. Ann. Indian Acad. Neurol. **16**(1), 26–33 (2013). https://doi.org/10.4103/0972-2327.107686
5. Othman, A.A., Tizhoosh, H.R.: Segmentation of breast ultrasound images using neural networks. In: Iliadis, L., Jayne, C. (eds.) AIAI/EANN -2011. IAICT, vol. 363, pp. 260–269. Springer, Heidelberg (2011). https://doi.org/10.1007/978-3-642-23957-1_30
6. Liu, S., et al.: Deep learning in medical ultrasound analysis: a review. Engineering **5**(2), 261–275 (2019). https://doi.org/10.1016/j.eng.2018.11.020
7. Huang, G., Liu, Z., van der Maaten, L., Weinberger, K.Q.: Densely connected convolutional networks. In: Proceedings - 30th IEEE Conference on Computer Vision and Pattern Recognition, CVPR 2017, 2017, vol. 2017-January, pp. 2261–2269. https://doi.org/10.1109/CVPR.2017.243
8. Vashishtha, V., Aju, D.: Nerve segmentation in ultrasound images. In: 2017 Innovations in Power and Advanced Computing Technologies, i-PACT 2017, vol. 2017-January, pp. 1–5 (2018). https://doi.org/10.1109/IPACT.2017.8244914
9. Rai, R., Sisodia, D.S.: Real-time data augmentation based transfer learning model for breast cancer diagnosis using histopathological images. In: Rizvanov, A.A., Singh, B.K., Ganasala, P. (eds.) Advances in Biomedical Engineering and Technology: Select Proceedings of ICBEST 2018, pp. 473–488. Springer, Singapore (2021). https://doi.org/10.1007/978-981-15-6329-4_39
10. Khobragade, V., Jain, N., Sisodia, D.S.: Deep transfer learning model for automated screening of cervical cancer cells using multi-cell images. In: Florez, H., Misra, S. (eds.) Applied Informatics: Third International Conference, ICAI 2020, Ota, Nigeria, October 29–31, 2020, Proceedings, pp. 409–419. Springer, Cham (2020). https://doi.org/10.1007/978-3-030-61702-8_28
11. Sobhaninia, Z., Emami, A., Karimi, N., Samavi, S.: Localization of Fetal Head in Ultrasound Images by Multiscale View and Deep Neural Networks, arXiv e-prints, p. arXiv:1911.00908 (November 2019)
12. Dhawan, A.P.: Image Segmentation, in Medical Image Analysis, 2nd Edition, Wiley-IEEE Press, pp. 1–6 (2010)
13. Kakade, A., Dumbali, J.: Identification of nerve in ultrasound images using u-net architecture. In: Proceedings - 2018 International Conference on Communication, Information and Computing Technology, ICCICT 2018, vol. 2018-January, no. ii, pp. 1–6 (2018). https://doi.org/10.1109/ICCICT.2018.8325894
14. "Ultrasound Nerve Segmentation | Kaggle (2016). https://www.kaggle.com/c/ultrasound-nerve-segmentation/data. Accessed 04 Aug 2019
15. Dey, S.: Hands-on Image Processing in Python, 1st ed., vol. 1. Packt Publishing Limited, Birmingham (2018)
16. Molau, S., Pitz, M., Ney, H.: Histogram based normalization in the acoustic feature space. In: 2001 IEEE Workshop on Automatic Speech Recognition and Understanding, ASRU 2001 - Conference Proceedings, pp. 21–24 (2001). https://doi.org/10.1109/ASRU.2001.1034579
17. Zelinsky, A.: Learning OpenCV—computer vision with the OpenCV library. IEEE Robot. Automat. Mag. **16**(3), 100 (2009)
18. Ojala, T., Pietikäinen, M., Mäenpää, T.: Gray scale and rotation invariant texture classification with local binary patterns. In: Vernon, D. (ed.) ECCV 2000. LNCS, vol. 1842, pp. 404–420. Springer, Heidelberg (2000). https://doi.org/10.1007/3-540-45054-8_27

19. Manikpuri, U., Yadav, Y.: Image enhancement through logarithmic transformation. Int. J. Innov. Res. Adv. Eng. (IJIRAE) **1**(8), 357–362 (2014)
20. Long, J., Shelhamer, E., Darrell, T.: Imperforated anus in calves and its surgical treatment. Intas Polivet **10**(2), 227–228 (2009)
21. He, K., Zhang, X., Ren, S., Sun, J.: Deep residual learning for image recognition. In: Proceedings of the IEEE Computer Society Conference on Computer Vision and Pattern Recognition, pp. 770–778 (2016). https://doi.org/10.1109/CVPR.2016.90
22. Anuragi, A., Sisodia, D.S.: Alcoholism detection using support vector machines and centered correntropy features of brain EEG signals (November 2017). https://doi.org/10.1109/ICICI.2017.8365294
23. Zhao, H., Shi, J., Qi, X., Wang, X., Jia, J.: Pyramid scene parsing network. In: Proceedings - 30th IEEE Conference on Computer Vision and Pattern Recognition, CVPR 2017, vol. 2017-January, pp. 6230–6239 (2017). https://doi.org/10.1109/CVPR.2017.660
24. Chaurasia, A., Culurciello, E.: LinkNet: exploiting encoder representations for efficient semantic segmentation. In: 2017 IEEE Visual Communications and Image Processing, VCIP 2017, vol. 2018-Januaary, pp. 1–4 (2018). https://doi.org/10.1109/VCIP.2017.8305148
25. He, K., Zhang, X., Ren, S., Sun, J.: Deep Residual Learning for Image Recognition (December 2015). http://arxiv.org/abs/1512.03385
26. Altman, D.G., Bland, J.M.: Statistics notes: diagnostic tests 1: sensitivity and specificity. BMJ **308**(6943), 1552–1552 (1994). https://doi.org/10.1136/bmj.308.6943.1552
27. Howard, A.G., et al.: MobileNets: Efficient Convolutional Neural Networks for Mobile Vision Applications (2017) http://arxiv.org/abs/1704.04861
28. Chaurasia, A., Culurciello, E.: LinkNet: Exploiting Encoder Representations for Efficient Semantic Segmentation (June 2017) https://doi.org/10.1109/VCIP.2017.8305148
29. Huang, G., Liu, Z., van der Maaten, L., Weinberger, K. Q.: Densely Connected Convolutional Networks (August 2016) http://arxiv.org/abs/1608.06993

Technical Losses (Tl) and Non-technical Losses (NTL) in Nigeria

Adeyemo Ayokunle[1], Sanjay Misra[1](✉), Jonathan Oluranti[1], and Ravin Ahuja[2]

[1] Center of ICT/ICE Research, Covenant University, Canaanland, Nigeria
Adeyemo.Dolapo@stu.cu.edu.ng, {Sanjay.misra, jonathan.oluranti}@covenantuniversity.edu.ng
[2] Shri Viswakarama Skill University, Gurgaon, India

Abstract. Generation of power by power stations passes through the complex and large network. Many types of equipment are used in the process of transmitting to the end-users, such as transformer, overhead lines, cables, and other equipment. It has been observed that the amount of power generated by Generating Companies (GENCO) compared to the amount of power reaching the Distribution Companies (DISCO) and neither does it matches the amount reaching the customers. This paper reviews all forms of Technical and Non-Technical losses and also lay emphasis on the accurate estimation on the amount of both losses recorded over the past years and how to reduce both Technical and Non-technical losses in Nigeria.

Keywords: Technical losses (TL) · Non-technical losses (NTL) · Distribution companies · Accurate estimation

1 Introduction

Analysis and reduction of losses is a very important aspect of the power distribution system, and it is a major determinant for the operation of a system with high efficiency. Energy loss occurs mostly when there is a need to distribute the network, the difference between GENCO and DISCO is the unit generated by either of them, and it is known as T&D losses. Estimate traditionally by means of energy balance have presented the total distribution of energy losses as the subtraction of energy consumed by the customer (billed and metered) from summated generated energy or distributed to feeders and substations.

$$T\&D\ Losses = (Energy\ Input\ to\ feeder\ (kwh)) - \frac{Billed\ Energy\ to\ Consumer\ (kwh)}{Energy\ Input\ kwh} \times 100$$

The power sector concerned with distribution is assumed to have the lowest amount in the power sector network because an estimated 50% can be attributed to distribution losses while transmission losses are around roughly 17% [1]. T&D losses are in two classes; the technical and non-technical [T&NT] (commercial) losses which account

L. Garg et al. (Eds.): ISMS 2020, LNNS 303, pp. 147–159, 2022.
https://doi.org/10.1007/978-3-030-86223-7_14

for the total energy losses, which makes it necessary for further analysis by estimating them individually. However, approximate methods are used for estimating technical losses, specifically for taking account of active power losses at severe load conditions by utilizing data processing models and documented data, followed by application of a loss factor to approximate power losses on an average and then by multiplication of average power losses in time (hrs) to derive the annual and monthly energy losses. The non-technical losses are then obtained by subtracting technical losses from total losses.

The present annual (technical) energy (MWh) losses on any distribution network can be calculated by the use of a formula, and this formula involves estimating the active power (MW) losses during the peak demand period (with, Fixed and Variable losses) and then, by reference to the distribution network's demand characteristics, i.e. the daily and seasonal variation in demand, to estimate associated, annual energy losses [2].

The LLF (Load Loss Factor) can be obtained using this formula:

$$LLF = \frac{\Sigma_{i=1}^{i=8760} Hourly\ demand\,(i)^2}{Maximum\ Demand^2 \times 8760}$$

The methodology used has great influence based on the intricate level of calculations applied to derive power losses at peak loss factors and also the properties of the distribution system along with the components associated with the analysis.

Eleven distribution bodies were unbundled from the defunct Power Holding Company of Nigeria (PHCN) and were privatized. Benin Electricity Distribution Company (BEDC) is one of the 11 distribution bodies with a very wide area of geographical spread: Delta, Edo, Ekiti, and Ondo, having a total of twenty-two business districts. Before the acquisition of BEDC, the body was the managing operative for the National Pre-Payment Metering Program (NPPMP) from 2007 to 2010. The major goal is to accomplish a significant reduction of technical and non-technical losses whilst enhancing the contentment of consumers by applying better customer service delivery.

2 Related Works

Smith [3] discussed how electricity theft has been on the increase in almost all the region of the world, which is enough reason why consumers' needs to be charged due to the low income realized from sales of electricity. Authors [3] categorize electricity theft as closely related to governance indicators that are countries without effective accountability are assumed to have more levels of theft, ineffectiveness in governance, political instability, and exceeding the level of corruption. It went further to talk about how it can be reduced by introducing tamper-proof meters, managerial inclusion on the job such as monitoring and regular inspection, and restructuring power systems regulation and ownerships in some cases (which are all technical solutions). Aguero [4] presents an improved power distribution systems efficiency by utilizing technologies, methodologies, and operational approaches, with rigid attention to the correct approximation and minimization of both technical and non-technical power losses. From the evaluation of losses point of view, which consists of efficient usage of data obtained from utility information systems, and making use of computational feeder models and dedicated simulation and modeling packages for accurate computation of both T&NT losses. The efficiency from the distribution setup can be greatly enhanced by reducing and controlling both the T&NT losses as predicted by [5] while developments in the industry such as the wide-spread of the smart grid concept provide up supplementary means to utilities by way of producing numerous options for the minimization of system losses and recent technology advances.

Davidson [6] talks about electrical power networks and the effectiveness of regulating non-technical losses within. He also further presents the effect of non-technical losses on the electricity industry and how it can be evaluated. The author was able to come up with help on how to reduce the power generation, the system capacity, and cost of power delivery.

Nagi et al. [7] also presented a new strategy for non-technical losses, the use of an innovative intelligence-based system, and also support vector machines (SVM) for analysis of electrical utilities. The result from the research shows the effectiveness of reducing NTL activities compared with the current actions taken by TNB. Fraud activities have increased over time; Nagi et al. [8] came up with model testing using an artificial intelligence-based technique as power detection utility, a historical kilowatt-hour data consumption in three towns in the Malaysian peninsular as a case study. The new fraud detection system built shows that the system is more effective in comparison with the existing ones.

Monedero et al. [9] highlighted a few methods of non-technical losses detection obtained from consumers using a power utility in Latin America and Spain: Endesa Company. The authors presented a means for identification of consumers with the peculiar downturn in the amount of energy consumed using windowed analysis alongside the Pearson coefficient.

Mwaura [10] talks about the adoption of a prepayment billing system as a means of reducing Non-Technical and energy losses in Uganda; in order to make this work, a study was developed to know the advantages of EPBS in the reduction of power theft, evaluating and comprehending how EPBS functions and its feasibility in Uganda.

Ramos [11] used a novel algorithm for selecting features using harmony search. Features like OPF, SVM-RBF, SVM-no Kernel, ANN-MLP, SOM, and k-NN have compared the latter features and concluded that the latter features are more accurate and effective in eliminating non-technical losses.

Viegas [12] identifies a topology that can be used for categorization of solutions in detecting non-technical losses; the likely attack and vulnerabilities are identified as well. The authors present loopholes in some existing techniques and the limitation such technique has. Aguero JR et al. [13] all presented a paper which reviews the operation approaches, technology, and methodologies focused on improving power distribution system, he also worked on the reduction of both TL and NTL.

Lewis [14] talked about how theft has cost the power industry so highly; he also evaluates the connection between electricity theft, power disruptions, and the resulting direct cost to an economy. Sahoo et al. [15] restructured the predictive model used in calculating loss for a branch in the energy distribution network by integrating to the network a temperature dependency of resistance. Mam et al. [17] discussed the distribution network reconfiguration based on BFOA (Bacteria foraging Optimization Algorithm) parallel with DNR (Backward-Forward Sweep) load flow method and GIS

(Geographical Information System). The author D. Suriyamongkol [18] talked about the analysis of loss-related statistics of Thailand's rural area power system market; also the magnitude and characteristics of electricity theft with utility company measured losses. A. J. Dick [19], Nizar et al. [20], and Antmann P [21] talked about both TL and NTL with the ways to reduce the losses.

Fourie et al. [22] paid attention to energy management research for the mitigation of electrical energy losses occurring in the electricity distribution network, and these losses are both TL and NTL. The model developed to measure the TL and NLT in the electrical distribution network is used to foretelling the technical energy losses, which enable them to develop a strategy to minimize the energy losses.

There are researches going on how to save energy, sustainability, and other related areas which we are not considering in detail [23–26].

3 Categories of Transmission and Distribution Losses

Transmission and distribution can be categorized in two classes:

I. Technical Losses.
II. Non-Technical Losses.

Technical Losses. Technical losses are caused majorly by the dissipation of energy in the equipment, type of conductors obtained for line transmission and sub-transmission, transformer, distribution line, and magnetic losses in transformers. A larger percentage of distribution lines in the primary and secondary stages while the sub-transmission and line transmission make up approximately 30% of the cumulative loss. It depends solely on the mode of operation and the characteristics of the network, which is about 25percent loss. While planning for the primary and secondary distribution, it's good to plan within limits.

Technical losses are of two types:

a) Permanent/Fixed Technical Losses: This type of loss takes the form of noise and heat, and it occurs during the time the transformer is energized. It does not vary with the current; the network effect of this comes in different variations like Open-Circuit, Dielectric, and Corona Losses, Losses from measuring and control elements because of continuous load, and also losses from current leakage. Fixed losses range within a third and a quarter of distribution network technical losses.
b) Variable Technical losses: Variable technical losses is directly proportional to the square of the current and vary with the amount of electricity distributed. Losses will fall by an increase in the trans-sectional area of cables and line for the load in question, which leads to an equivalent exchange between the cost of capital expenditure and the cost of losses. Variable losses are range from two-thirds to three-quarters responsibility of distribution network technical losses. It advised that the more favorable average consumption rate on a distribution network that takes losses cost during design into consideration should fall under 30%. Such losses include; contact resistance losses, joule and impedance losses in lines in each voltage level.

Major Causes for Technical Losses: a. Elongated Distribution lines: High I2 R losses in the line and high line resistance are resulted due to the primary and secondary distribution lines in underdeveloped regions covering lengthy distances to distribute loads dispersed over massive areas. They are practically 11 kV and 0.415 kV lines. Disorganized expansion of sub- distribution systems and sub-transmission into more recent areas, extensive rural supply of electricity via excessively long 11 kV and LT lines.

b. Insubstantial capacity of Conductors of Distribution lines: The conductor size of feeders has to be adequate. Rural loads normally are disorganized and customarily fed by radial feeders, but the conductors' specifications should be chosen based on a KVA x KM capability of accepted conductors convenient to specific voltage regulations.

c. Low Power Factor(LPF): LPF supplements large distribution losses. For any load in question, should there be an LPF, the rate of current usage will be high, and the square of the proportionality of the square of that current to the losses will be higher; however, losses in the line caused by trivial PF can be minimized by the application of shunt capacitors to improve the Power Factor. These shunts capacitors can be affixed on the secondary side (11 kV side) of the 132 kV/11 kV power transformers or alternatively, at distinct positions across the distribution Line.
The distribution transformer is preferably located at the load center to ensure the voltage drop stays in the range of acceptable boundaries to ensure adequate minimization of the voltage drop across the line to remote consumers. More often, distribution transformers are not situated axially in conjunction with consumers. Tentatively, consumers that are very far off will not receive a good voltage level despite the optimal operation of the transformer, and this inconsequentially leads to higher line losses, which occur due to decreased voltage at the consumers' end.

d. Effect of Load Factor (LF) on Losses: The amount of usage of power by consumers differ over different days and seasons. Commercial consumer load customarily peaks in the early afternoon, while residential consumers customarily reach their peak load demand around the evening. As a load, a primary reason behind losses, maintaining power utilization all through the day will reduce the peak power loss and also overall energy losses.

$$Load\ Factor = \frac{Average\ load\ in\ a\ specified\ time\ period}{peak\ load\ during\ that\ time\ period}$$

The variation of the load is termed Load Factor (LF), and it always ranges between 0 and 1.

e. Poor Craftsmanship: Poor Craftsmanship also offers up its own significant contribution to the increase of distribution losses. Joints for example, joint tend to be sources of power loss, so the amount of joints used must be as minimal as possible and efficient techniques must be applied in connecting joints to ensure rigidly a rigid network.

f. 3-Phase Load Balancing: 3-Phase load balancing repeatedly and systematically across a network potentially minimizes losses drastically. It can be executed rather simply on overhead networks, and this, therefore, leads to cost-effective loss minimization, provided there are fitting motivations.

Other Reasons for Technical Losses:

I) Transformer Sizing and selection.
II) Feeder Phase Current and Load Balancing.
III) Switching off Transformers.
IV) Leakages and power loss.
V) Line Overloading.
VI) Enormous pull on the current by induction loads, which leads to the reduced voltage at customer terminals.
VII) Uneven load distribution between 3-phases in L.T system, which leads to enormous neutral currents.
VIII) Inferior grade of equipment utilized for water provision and irrigation purposes in underdeveloped regions and industrial load in developed regions.

Non-Technical losses. The cumulative measure of T&D losses, T&ND losses, results in the distinction between the total electric energy injected into the T&D and the one coupled with the revenue acquired from bills paid by consumers. Whereas technical losses make up the electric energy discharged through the devices necessary to execute the T&D of electricity, the NTLs make up the difference between total T&D losses and technical losses. NTLs can be approximated, but definite quantification is impossible. NTLs have been established to be majorly a direct outcome of fraudulent activities such as hardware tampering, line tapping theft, and bills gone unpaid. Also, inconsistencies in the quantification of consumption/billing and conspiracy involving utility staff also make up a source of NTLs [10].

Major Causes of Non-Technical Losses:

a. Power Theft: The most common type of NTL is power theft, and it involves the energy supplied to consumers that goes unaccounted for in the consumer's energy meter. Consumers tamper with meters through mechanical manipulations, affixation of very powerful magnets producing disturbance in the disc rotation with unfamiliar elements, and design of remote control that hinder optimal meter performance, etcetera.
b. Metering Inaccuracies: Various meters produce a measurable amount of problems that necessitate the ideal standards to be created. The distinction between the volumes of energy supplied through the meters and the volumes the meter registers are the Losses due to metering inaccuracies. Meters are required to range from + 2.5% and −3.5% accuracy while aged meters start operation with trivial errors; as they age, they tend to present results in under-recording. Since the advent of electronic meters, there have been advanced minimizations in error readings of meter.
c. Unmetered Losses for Very small load: This occurs because the load is quite minute and installation of energy meters is not cost-effective at all and unfeasible; a scenario involving this is the streetlight.
d. Unmetered supply: Unaccountable provided power to the agricultural pump is the main cause for financial losses; the agronomical tariff depends on the unit horsepower (H.P.) of the rotors. This sort of load gets accredited at the low load resolution.

e. Meter Reading Error: All electrical energy measurements must be done with calibrated Meters, and energy meters with deficiencies must be substituted with functioning meters promptly. The major cause for faulty meters are untraceable meters, meter burning, heavy load due to terminal box burn out of meter, inappropriate testing and calibration of meter, C.T.radio error and reducing the recording, Meter number mismatch, meter not read.
f. Billing Problem: Complaints regarding billings of the customer and under billing, late bill receipt, getting the inaccurate bill, inaccurate accounting of meter, misguided tariff, inaccurate computation, and estimated bill.

4 Electricity as an industry in Nigeria

Electric utilities are no more monopolistic enterprises in Nigeria with 11 Distribution companies. Regulatory mandates, retail competition, rising customer aspirations, fast obsolescence, changing requirements of metering, tariff, and billing, have collectively contributed to a fundamental shift in customer service, Reforms, restructuring, regulation, and deregulation have ordained a marked degree of acumen and professionalism in this industry. Cost intensiveness involves wide equity and debt financing base, investors and creditors always looking forward to profitability and returns. The distribution industry has become complex and demanding; cost-effectiveness is the key to remain in business with a focus on Cost-effectiveness with no dilution in efficiency, improved cash flow, Technical and man-made loss control, Profitability with EPS, and a positive revenue stream will make this sector bankable.

4.1 Power Sector Growth in Nigeria

The power sector in Nigeria has grown many manifolds, but the gap between supply demands is widening. Peaking shortages are all time of the year. Despite the expansion of distribution sectors and sub-transmission, investments haven't matched the capability extension. A huge portion of the populace don't have access to the electricity supply. In the state control non-competitive setup, consumer contentment is ignored, and attention is not paid to revenue creation. Investments require a commercial approach to stimulate people towards them.

4.2 Verified Losses and Loss Reduction Benchmarks

Table 1 below summarizes the verified losses for all the discos, the initial Figs. [1, 2] as specified in the privatization RFPs and the yearly loss reduction trajectory that has been committed to by each winning bidder.

4.3 Initiation of Loss Minimization

Disco is designed to reduce the losses each year based on the loss reduction trajectory committed by their new core investors promised in their bid documents. In calculating the aggregate technical, commercial losses for the Disco, and in accordance with the

Table 1. Verified losses

Disco	Bid basis	Baseline verified	1st Year	2nd Year	3rd Year	4th Year
Abuja	35	52.77	36.41	26.58	24.72	22.03
Benin	40	54.2	44.44	35.56	27.56	20.67
Enugu	35	59.1	58.73	44.05	31.71	20.3
Ibadan	35	42.6	35.78	29.09	23.15	18.56
Jos	40	58	52.03	47.34	21.45	34.63
Kaduna	40	48.4	32.43	20.43	12.67	7.83
Kano	40	48.4	41.4	31.87	23.91	18.65
Eko	35	29.4	21.46	15.88	12.55	11.22
Ikeja	35	32.93	24.7	18.05	12.83	10.45
P/H	35	55.04	46.78	38.6	30.88	24.7
Yola	40	57.6	43.2	36.23	31.22	27.46

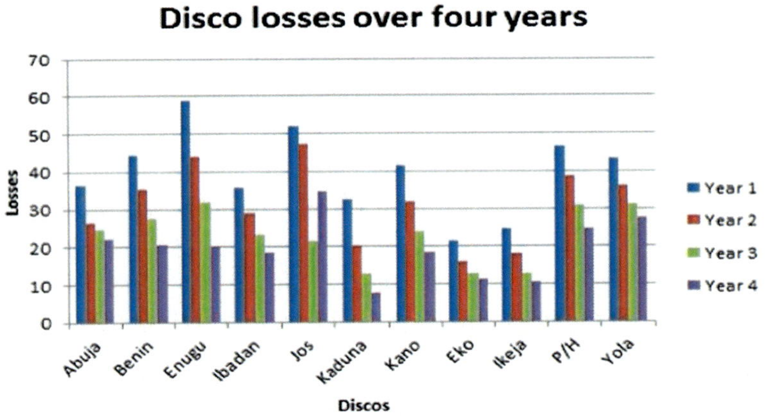

Fig. 1. Disco losses obsered in 4 years

privatization transaction documents and the extant Tariff Order. The commission adopted 1st Nov 2012 to 31st Oct 2013 as the base year and began to implement individual projections from Disco's loss minimization from 1st Nov 2013, the period on which Discos were formally handed over to their respective core investors.

It should be noted that the loss reduction has no direct correlation with the declaration of TEM. In other words, the disco is obliged to commence the loss minimization instantly after taking over the proprietorship and control of these companies. However, in order to synchronize Disco's annual loss reduction trajectory with the new tariff year has been extended by 2 months.

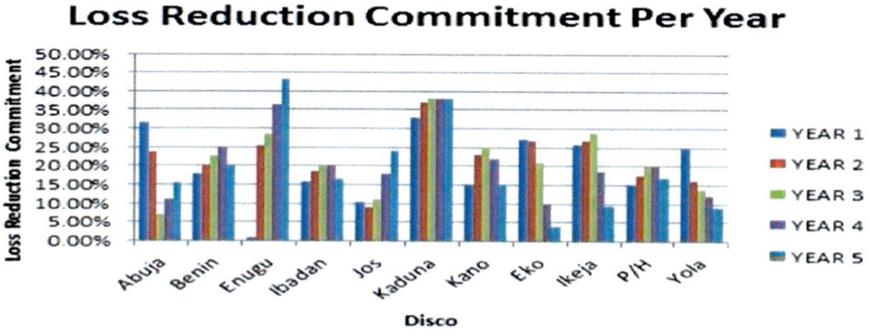

Fig. 2. Loss reduction commitment per Year

5 Reduction of Losses

This section entails a discussion of how to reduce Technical and Non-Technical Losses.

5.1 Process to achieve minimization of technical losses

a) Transforming LV to HV Line:
High voltage feeders in towns are surrounded by Low Voltage (430 V) in many distribution pockets; at lesser voltages, more current from the conductor flows from the same power input, producing higher I2 R losses. The transformation of aged LV (430 V) feeders to HV is quite costly and most times not financially advisable but in a situation where the LV (430 V) primary feeders are of proper working condition, various step-down power transformers can be implemented at the verges of the 430 V area that will incite reduction of copper losses by infusing load current at numerous positions (i.e. reduction of the cumulative conductor current and the distance covered by the current to reach the load).

b) Industrial Consumers receive line directly from feeder:
The system is implemented uniquely so as to ensure the consumer receives enormous from feeder to direct power line.

c) Utilizing High Voltage Distribution Service (HVDS) for Agro-Consumers:
In HVDS, a collection of people/groups of agro-consumers like 2 or 3 are fed a direct 11 kV line per agricultural pump set and makes use of 15 KVA for giving each of these collections through small-scale. The losses are minimized severely because of the use of the minimal length of distribution line, a good standard of power supply without a drop in voltage, reduced motor burnout courtesy of minimal voltage variations, and a high standard of power to prevent transformer overloading.

d) Utilization of an Arial Bundle Conductor (ABC): The use of an ABC to subsidize line defects when LT line is not totally avoidable, to prevent line tampering.

e) Reduction of the amount of Transformers: Transformers are solely the cause of approximately 50% of network losses. Reduce the number of transformation steps, and the loss minimization effect on Distribution lines can be achieved with high-efficiency distribution transformers.

f) Average Capacity use of Feeders: Losses in distribution lines will escalate by overloading the Distribution Feeder. The drag on the distribution line is directly proportional to variable losses, and the ideal average use rate of network distribution wirings is preferably supposed to be at 30% if the expenditure caused by losses is accounted for as suggested.
g) Substitution of Aged Conductor/Cables: By ensuring the use of a larger trans-section area of the conductor, minimization of losses occur; however, the price to pay will be enormous, so by adequately planning the future load, an ideal equilibrium should be maintained between losses in the network and cost of investments.
h) Improvement Program/Feeder Renovation: Identification of the least reinforced regions in the distribution system and then reinforcing these regions, implementation of distribution transformer that operates at lower capacity at each customer's surroundings and not the collection formation and swapping of distribution transformers with the ones whose load is lower and losses minimal like the amorphous core transformers and utilization of single-phase transformers to supply sedentary and non-sedentary loads in underdeveloped areas.

5.2 Process to achieve minimization of Non-Technical Losses

a) Mapping of Distribution Line Data: Using specifications such as the size of conductor lengths of line, etcetera, in the mapping of the complete primary and secondary distribution system, gathering of data as regards existing loads, operating conditions, adequate planning of future loads etc. Establishment of long-standing strategies for phased reinforcement and enhancement of the distribution systems and also transmission system.
b) Incorporation of energy audits strategies: It should be made mandatory that specific organizations and utilities that fall under a certain measurable category conduct energy audits of their setup.
c) Regulation of power theft by drives installed for power theft checking; One of the main challenge associated with all-electric utilities (DISCOS) in Nigeria is theft, and the effect of theft is not limited bounded by the losing income, the standard of power in load shedding, low voltage, and voltage dips are also affected. A proper seal must be installed at each terminal box of the meter at the CT/PT terminal in order to avert theft of power.
d) Replacement of Faulty/Sluggish Energy Meter: The substitution of defective meters by the Distribution Agency is necessary so as to reduce unmetered electrical energy and testing of meters to make sure it is working properly. The substitution of aged meters with precise Electrostatic meter for precise and concise measurement of energy usage. The utilization of properly and tightly sealed boxes to house meters to ensure the meters cannot be meddled with.
e) Facility for Bill Collection: Creation of more and standard bill payment chambers, and also dropbox facilities in all regions that handle the collection of payments. E-payment can be used as means of payment because it gives relief to customers, and Discos will get fast and regular payment through it. A customer who does not pay bills should be disconnected.

f) Reduction of Sub Division of Debit areas: Debt collection from defaulters through various ways such as legal correspondence and judicial litigations also making sure involvement of police when necessary is followed for disconnection of any consumer found wanting of required appeasements.
g) Watchdog effect on users: It should be made known in very clear terms to the users that the distribution agency completely oversees and keep track of all activities and consumption as seen fit, which enables the body very fast knowledge of any irregular consumption achieved by meddling with a meter and allows them to take any corrective and punishable action as required and seen fit. The customers stealing power will stop once they are well-versed in the utility's ability to detect and register it; sufficient penalty must be given to such individuals.

6 Conclusion

The efficiency of a distribution system can be enhanced significantly by reducing and controlling both T&NT losses. The major supplement and degree of system losses differ in more developed nations. Impressively, NT losses is not such a unique challenge faced with only under-developed nations such as Nigeria. Checking the necessity of electrical power worldwide and notable energy consumption and rates, comparatively, minimal system losses in a developed country might possess an identical impeccable financial value than the case of enormous system losses in an underdeveloped nation. Recent advance in technology and innovative practices like the utilization of the smart grid concept benefits utilities with supplementary tools and strong potential options for minimizing system losses. The evaluation of all types of losses was done for both Technical and Non-Technical losses, and the losses in Nigeria over the past four years were evaluated. The methods that can be employed to deal with such losses were evaluated in this paper.

Acknowledgment. The authors appreciate the sponsorship from Covenant University through its Centre for Research, Innovation and Discovery, Covenant University, Ota Nigeria.

References

1. Parmar, J.: Total losses in power distribution and transmission lines (2013). Accessed 19 Aug 2021. https://electrical-engineering-portal.com/total-losses-in-power-distribution-and-transmission-lines-1
2. Ekwue, A.O.: On the correctness of load loss factor. Niger. J. Technol. **34**(3), 546–547 (2015)
3. Smith, T.B.: Electricity theft: a comparative analysis. Energy Policy **32**(18), 2067–2076 (2004)
4. Aguero, J.R.: Improving the efficiency of power distribution systems through technical and non- losses reduction. In: IEEE PES Transmission Distribution Conference Exposition, pp. 1–8 (2012)
5. Davidson, I.E.: Evaluation and effective management of non-technical losses in electrical power networks. In: IEEE AFRICON, 6th Africon Conference Africa, pp. 473–477, June 2002

6. Nagi, J., Mohammad, A., et al.: Non-technical loss analysis for detection of electricity theft using support vector machines. In: IEEE International Conference Power Energy, PECon 2008, pp. 907–912 (2008)
7. Nagi, J., Mohammad, A.M., et al.: Non-technical loss detection for metered consumers in power utility using support vector machines. IEEE Trans. Power Deliv. **25**(2), 1162–1171 (2010)
8. Monedero, I., et al.: Detection of frauds and other non-technical losses in a power utility using Pearson coefficient, Bayesian networks and decision trees. Int. J. Electr. Power Energy Syst. **34**(1), 90–98 (2012)
9. Mwaura, F.M.: Adopting electricity prepayment billing system to reduce non-technical energy losses in Uganda: lesson from Rwanda. Util. Policy **23**, 72–79 (2012)
10. Ramos, C.C.O., et al.: A novel algorithm for feature selection using harmony search and its application for non-technical losses detection. Comput. Electr. Eng. **37**(6), 886–894 (2011)
11. Viegas, J.L., et al.: Solutions for detection of non-technical losses in the electricity grid: a review. Renew. Sustain. Energy Rev. **80**, 1256–1268 (2017)
12. Aguero, J.R.: Improving the efficiency of power distribution systems through technical and non-technical losses reduction. In: Proceedings of the IEEE PES transmission and distribution conference and exposition (2012)
13. Lewis, F.B.: Costly throw-ups: electricity theft and power disruptions. Electr. J. **28**(7), 118–135 (2015)
14. Sahoo, S., et al.: Electricity theft detection using smart meter data. In: 2015 IEEE Power Energy Society Innovative Smart Grid Technologies Conference, pp. 1–5 (2015)
15. Gonen, T.: Electric Power Distribution System Engineering. McGraw- Hill Inc. (1986)
16. Mam, M., Leena, G., Saxena, N.S.: Distribution network reconfiguration for power loss minimization using bacterial foraging optimization algorithm. Int. J. Eng. Manuf. **6**(2), 18–32 (2016)
17. Suriyamongkol, D.: Non-technical losses in electrical power system. MSc thesis, Ohio University, November 2002
18. Dick, A.J.: Theft of electricity—how UK electricity companies detect and deter. In European Convention on Security and Detection, pp. 90–95, Brighton, U.K, 16–18 May 1995
19. Nizar, A.H., et al.: A data mining based NTL analysis method. In: Proceedings of the 2007 IEEE Power Engineering Society General Meeting, No. 3, pp. 1–8 (2007)
20. Antmann, P.: Reducing technical and non-technical losses in the power sector background paper for the World Bank group energy sector strategy. Technical reports (2009)
21. Fourie, J.W, Calmeyer, J.E.: A statistical method to minimize electrical energy losses in a local electricity distribution network. In: Proceedings of the 7th IEEE AFRICON Conference Africa: Technology Innovation, Gaborone, Botswana September 2004
22. Adesanya, A., Misra, S., Maskeliunas, R., Damasevicius, R.: Prospects of ocean-based renewable energy for West Africa's sustainable energy future. Smart Sustain. Built Environ. **10**(1), 37–50 (2020)
23. Ayodele, E., Misra, S., Damasevicius, R., Maskeliunas, R.: Hybrid microgrid for microfinance institutions in rural areas–a field demonstration in West Africa. Sustain. Energy Technol. Assess. **35**, 89–97 (2019)
24. Ogunleye, O., Alabi, A., Misra, S., Adewumi, A., Ahuja, R., Damasevicius, R.: Comparative study of the electrical energy consumption and cost for a residential building on fully AC loads vis-a-vis one on fully DC loads. In: Jain, V., Chaudhary, G., Taplamacioglu, M.C., Agarwal, M.S. (eds.) Advances in Data Sciences, Security and Applications. LNEE, vol. 612, pp. 395–405. Springer, Singapore (2020). https://doi.org/10.1007/978-981-15-0372-6_32

25. Aderemi, O., Misra, S., Ahuja, R.: Energy consumption forecast using demographic data approach with Canaanland as case study. In: Bhattacharyya, P., Sastry, H.G., Marriboyina, V., Sharma, R. (eds.) Smart and Innovative Trends in Next Generation Computing Technologies, pp. 641–652. Springer Singapore, Singapore (2018). https://doi.org/10.1007/978-981-10-8657-1_49
26. Adeyemi-Kayode, T.M., Misra, S., Damaševičius, R.: Impact analysis of renewable energy based generation in West Africa–a case study of Nigeria. PROBLEMY EKOROZWOJU–Probl. Sustain. Dev. **16**(1), 67–78 (2021)

Kinetic Gas Molecular Optimized (KGMO) Artificial Neural Network (ANN) Based Software Reliability Prediction for Banking Applications

P. Lakshminarayana[1] and T. V. Suresh Kumar[1,2](✉)

[1] B.M.S. College of Engineering, Bangalore, Karnataka, India
[2] M.S.R Institute of Technology, Bangalore, Karnataka, India

Abstract. Software Reliability is one of the major considerations in the banking sector to improve security and quality. Rapid changes in hardware and software technologies lead to creations of new techniques and necessities creating and approving the dependability foreseeing model for every strategy. In this work, the applicability of Kinetic Gas Molecule Optimization (KGMO) with Artificial Neural Network (KGMO-ANN) as a model to predict the software Reliability. The parameter of the learning process of ANN can be adjusted by using the KGMO optimization process. The model has been applied on data sets collected across several standard software projects during system testing phase with fault removal To evaluate the performance technique, Root Mean Square Error (RMSE), Average Error (AE), Mean Absolute Error (MAE), Mean absolute percent error (MAPE) and Normalized Mean Square Error (NRMSE) will be calculated. From various evaluation technique, found that high accuracy is obtained from proposed Reliability Prediction technique.

Keywords: Reliability prediction · Kinetic Gas Molecule Optimization · Artificial Neural Network · Root Mean Square Error · Normalized Mean Square Error

1 Introduction

The reliability of a software system relies on the number of undesirable defects. The software has become the dynamic quantity of most of the complex applications and individuals are working under the fast or secondary influence of software. In this manner, it is very essential to guarantee the reliability of the software system. The presentation of the deficiency expectation model relies upon the features that are utilized to prepare it. Redundant and irrelevant features can ruin the presentation of an order model. The software development life cycle incorporates various stages, for example, prerequisite analysis, design, construction, testing, deployment, and maintenance, typically bolstered by different exercises, for example, configuration management, engineering management, and quality assurance. Software technology's current software sizes have become larger and additional software functions must be reached to satisfy client requirements.

L. Garg et al. (Eds.): ISMS 2020, LNNS 303, pp. 160–170, 2022.
https://doi.org/10.1007/978-3-030-86223-7_15

Software reliability is a significant factor in software quality. Software reliability is the probability that software won't bring on any disappointment of a framework for a predetermined period under the predefined conditions. In this work, the applicability of Kinetic Gas Molecule Optimization (KGMO) with Artificial Neural Network (KGMO-ANN) as a model to predict the software Reliability. The parameter of the learning process of ANN can be adjusted by using the KGMO optimization process, Root Mean Square Error (RMSE), Average Error (AE), Mean Absolute Error (MAE), Mean absolute percent error (MAPE) and Normalized Mean Square Error (NRMSE) used in this technique. Software reliability assumes a key role in software quality. In request to improve precision and consistency of software reliability prediction, in this paper, we have proposed the relevance of Kinetic Gas Molecular Optimized (KGMO) Artificial Neural Network (ANN) Based Software Reliability Prediction for Banking Applications.

Software fault prediction (SFP) is typically used to predict faults in software components. Solve this problem, we have presented an adaptive technique that combines Kinetic Gas Molecular Optimized (KGMO) Artificial Neural Network (ANN) Based Software Reliability Prediction. Component-Based Software (CBS) is an ongoing methodology in the field of software engineering that centres on totalling components into complex software systems with the quick advancement of component innovation. This methodology gives a few favourable circumstances, for example, productivity, quality, reusability, reduces maintenance overheads, and time-to-market. The reliability can be anticipated by figuring the reliability of components separately and the interconnection methodology between components. Reliability forecasting of CBS includes disappointment forecasting methods that evaluate framework reliability quantitatively.

The remainder of this paper is organized as follows: Sect. 2 gives prior work which is performed in the software reliability prediction. Section 3 explains the proposed method of SRP-KGMO-ANN in detail. Section 3 discusses the experimental results to demonstrate the performance and characteristics. Section 4 The parameters are used in this work of the proposed method, and lastly, Sect. 5 Results and Discussion of this paper.

2 Literature Survey

A great deal of works is introduced to anticipate and appraise the software reliability; some of them are inspected here, Jin et al. carried out an estimation of distribution algorithms (EDA) to maintain the diversity of the population, and then a hybrid improved estimation of distribution algorithms (IEDA) and SVR model, called IEDASVR model. IEDA was used to optimize parameters of SVR, and the IEDA-SVR model was used to predict software reliability. They compared the IEDA-SVR model with other software reliability models using real software failure datasets.

Arunima Jaiswal et al. demonstrated the ML techniques for software reliability prediction based on selected performance criteria. They were applied ML techniques including, feed-forward back propagation neural network, general regression neural network, support vector machines, multilayer perceptron, Bagging, cascading forward back propagation neural network, instance-based learning, linear regression, M5P, reduced error pruning a tree, M5Rules to predict the software reliability on various datasets being chosen from industrial software. Based on the experiments conducted, it was observed that yields better results in all the cases and thus can be used for predicting software reliability since it predicts the reliability more accurately and precisely as compared to all other above-mentioned technique.

Mengmeng Zhu et al. carried out a non-homogeneous Poisson process (NHPP) software reliability model with a pioneering idea by considering software fault dependency and imperfect fault removal. Two types of software faults were defined, Type I (independent) fault and Type II (dependent) fault, according to fault dependency. Two phases of debugging processes, Phase I and Phase II, were proposed according to the debugged software fault type. A small portion of software faults that software testers were not able to remove was also considered in both phases in the proposed model. The illustration of the model's effectiveness was based on the three datasets collected from industries.

Chander Diwaker et al. advocated a Bio-Inspired Soft Computing Techniques for predicting software reliability. Their paper focused on the assessment of commonly used soft computing technique like Genetic Algorithm (GA), Neural-Network (NN), Fuzzy Logic, Support Vector Machine (SVM), Ant Colony Optimization (ACO), Particle Swarm Optimization (PSO), and Artificial Bee Colony (ABC). They represent the working of soft computing techniques and the assessment of soft computing techniques to predict reliability. The parameter considered while estimating and prediction of reliability was also discussed.

3 SRP-KGMO-ANN

In this work, the applicability of Kinetic Gas Molecule Optimization (KGMO) with Artificial Neural Network (KGMO-ANN) as a model to predict the software Reliability. as a model to anticipate the number of deficiencies in the software during advancement and testing forms. The model has been applied to data sets gathered over a few standard software projects during the framework testing stage with flaw expulsion. Banking sector data is utilized for the testing process.

3.1 Block Diagram

The Block diagram of proposed software reliability prediction can be shown in Fig. 1. here bug data will be collected from the database. The standard database will be used to perform the training and testing process. While doing training. The standard target value will be assigned based on the real software reliability value which is stored in the database. All the input data will be normalized to 0 to 1 to increase the training performance. After that, the data will be arranged properly to perform the training process further. In the training section, the weight of the ANN will be optimized by using the KGMO algorithm. The minimum mean square error will be considered as The objective function for KGMO operation. Once the training performed the trained neural network will be stored to perform the testing operation as shown in Fig. 1.

3.2 Kinetic Gas Molecular Optimized (KGMO)

Populace based optimization algorithms follow some looking through strategies that merge to the ideal arrangement according to the target work where it begins from the random position of the plausible district. KGMO optimization is developed by the motivation of Kinetic molecular theory. According to the atomic theory of gas each article is made out of enormous volumes of atoms or molecules. Fundamentally, all the properties of the gases are depending upon the actions of the molecules that form the gas.

In the Kinetic Gas Molecular optimization algorithm, the gas molecules are the operators in search space. Kinetic energy is used in assessing performance. The gas particles move in the holder until they converge in the bit of the compartment that has the most reduced temperature and kinetic energy.

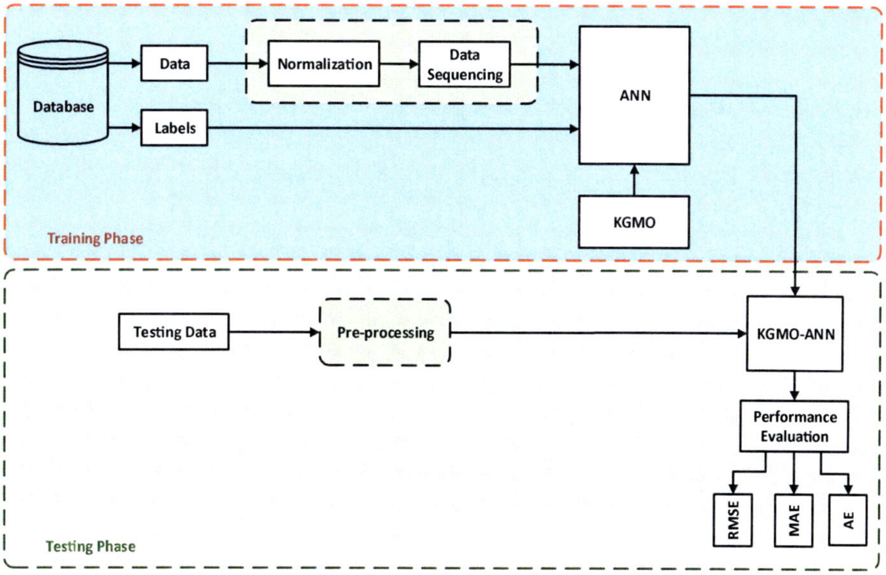

Fig. 1. Block diagram of proposed software reliability prediction

It is understood that gas atoms attract each other subject to feeble electrical intermolecular Van Der Waal powers, where the electrical force is the delayed consequence of positive and negative charges in the particles. In the KGMO, every gas atom (operator) has four subtleties: position, kinetic energy, speed, and mass. The kinetic energy of every gas atom chooses its speed and position.

In the algorithm, the gas atoms investigate the entire search space to arrive at the point that has the most reduced temperature. Next, consider a framework that has N operators (gas atoms). The position of the mth operator is characterized by

$$P_m = P_m^1, \ldots P_m^c \ldots, P_m^z for\,(i = 1, 2, \ldots .K) \quad (1)$$

Where P mth represents the position of the mth agent in the cth dimension.

The velocity of the ith agent is presented by

$$\upsilon_m = \upsilon^1_{m}, \ldots \upsilon^c_m \ldots, \upsilon^z_m for\ (m = 1, 2, \ldots .K) \tag{2}$$

where υ^c_m represents the velocity of the mth agent in the cth dimension.

The kinetic energy, in turn, is defined as

$$KE^c_m(u) = {}^3\!/_2 \left(K_y\, U^c_m(u), j_m = \left(j^1_m, ..j^c_m, ..., j^z_m\right), for\ i = 1, 2, 3.... z\right) \tag{3}$$

where N is the number of gas molecules, y is the Boltzmann constant, and $U^c_m(u)$ is the temperature of the ith agent in the cth dimension at time u. The velocity of the molecule is updated by

$$\begin{aligned} \upsilon^c_m(u+1) &= U^c_m(u)\omega v^c_m(u) + l1r(u)\left(e^c_{best} - A^c_m(u)\right) \\ &+ l2r(u)\left(\left(\frac{\left(F^c_{best}(u) + F^c_{best}(u-1) + F^c_{best}(U-2)\right)}{K_E}\right) - A^c_m(u)\right) \end{aligned} \tag{4}$$

where U^c_m for the converging molecules reduces exponentially over time and is calculated as

$$U^c_m = 0.95A\left(U^c_m(u-1)\right) \tag{5}$$

The vector $Fbest_m = \left(Fbest^1_m, Fbest^2_m, \ldots ..Fbest^z_m,\right)$ represents the best past position of the ith gas atom, and $ebest_m = \left(ebest^1_m, ebest^2_m, \ldots ..ebest^z_m,\right)$ is the best past position among the entirety of the particles in the holder. K represents the complete number of past best taken for averaging. The speed and the situation of every molecule are introduced by arbitrary vectors inside the relating ranges. Here, $[-\upsilon min, \upsilon max]$ is utilized as the cutoff points of the gas atoms' speed. [13] On the off chance that $|vi| > Vmax$, at that point $|vi| = Vmax$. w is the idleness weight that mirrors the gas atom's protection from moderate its development. Furthermore, r(t) is a uniform arbitrary variable in the interim [0,1] at time u, which is utilized to give a randomized trademark to the search algorithm. l1 and l2 are two speeding up constants.

The base fitness function is found by utilizing

$$\begin{aligned} F_{best} &= d(a_m), ifd(a_m) < d(Fbest_m) \\ e_{best} &= d(a_m), ifd(a_m) < d(ebest_m) \end{aligned} \tag{6}$$

3.3 Artificial Neural Network (ANN)

Artificial Neural Networks are computing systems that are inspired by biological neural networks. Such systems learn to perform tasks by considering examples. They might learn to identify images that contain information by analyzing from the examples.

ANN are multi-layer fully-connected neural nets that look like the figure below. They consist of an input layer, multiple hidden layers, and an output layer. Every node in one layer is connected to every other node in the next layer. We make the network deeper by increasing the number of hidden layers. It is a non-linear and adaptive system that learns to perform functions from data. ANN is said to be adaptive because it can change its parameters when the parameters of the system change in its training phase. After training the parameters of the network are fixed then this network is used to solve the problems. There are two modes of operation one is the feed-forward mode and another one is backpropagation mode. In this paper the Implemented.

ANN has one input and one output layer with one or two hidden layers. In the backpropagation method, every time the output is compared with the target, and the error is computed. The weight of the ANN is adjusted to reduce the error every time the weight changes the neural model gets closer to the targeted output. The excellent neural network training method used in the proposed system is the conjugate gradient backpropagation algorithm.

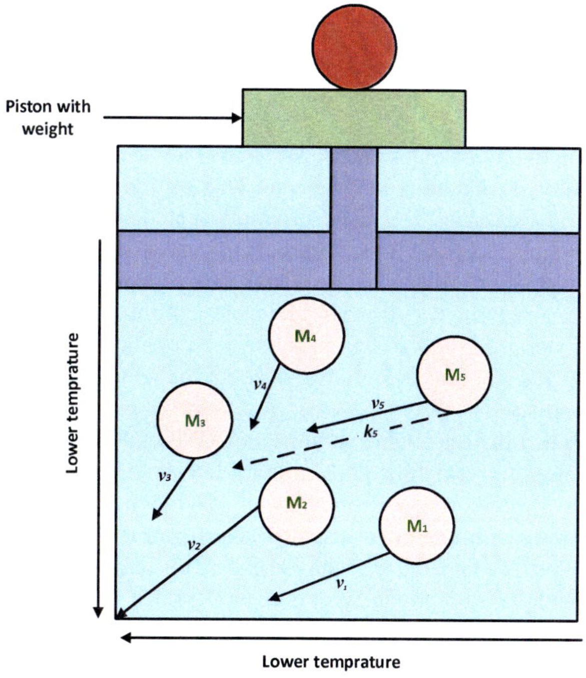

Fig. 2. KGMO algorithm, practical model

A given node takes the weighted sum of its inputs and passes it through a non-linear activation function. This is the output of the node, which then becomes the input of another node in the next layer. The signal flows from left to right, and the final output is calculated by performing this procedure for all the nodes. Training this deep neural network means learning the weights associated with all the edges.

The equation for a given node looks as follows. The weighted sum of its inputs passed through a non-linear activation function. It can be represented as a vector dot product, where n is the number of inputs for the node.

$$z = f(x.y) = f\left(\sum_{i=1}^{n} x_i y_i\right) \tag{7}$$

$$x \in d_{1\times n}, y \in d_{n\times 1}, z \in d_{1\times 1} \tag{8}$$

I omitted the bias term for simplicity. Bias is an input to all the nodes and always has the value 1. It allows to shift the result of the activation function to the left or right. It also helps the model to train when all the input features are 0. If this sounds complicated right now you can safely ignore the bias terms. For completeness, the above equation looks as follows with the bias included.

$$z = f(b + x.y) = f\left(b + \sum_{i=1}^{n} x_i y_i\right) \tag{9}$$

$$x \in d_{1\times n}, y \in d_{n\times 1}, b \in d_{1\times 1}, z \in d_{1\times 1} \tag{10}$$

4 Results and Discussion

An adaptive technique that combined KGMO and ANN technique has been proposed to evaluate the software reliability in this paper. This section gives detailed information about the result and discussion. This project is implemented using MATLAB 2018a. This work is carried out using a PC with an i5 processor and 4 GB RAM. The performance of the proposed technique is evaluated and compared with the conventional techniques such as the applicability of Kinetic Gas Molecule Optimization (KGMO) with Artificial Neural Network (KGMO-ANN) as a model to predict the software Reliability. The parameter of the learning process of ANN can be adjusted by using the KGMO optimization process.

Table1 shows that there are some parameters KGMO-ANN, PSO-ANN, GA-ANN, ANN which are used to predict the faults. Figure 1 shows the block diagram for SRP-KGMO-ANN. Figure 2 Shows the Artificial Neural Network (ANN) ANN has one input and one output layer with one or two hidden layers. In the backpropagation method, every time the output is compared with the target, and the error is computed. To evaluate the proposed technique, Root Mean Square Error (RMSE), Average Error (AE), Mean Absolute Error (MAE), Mean absolute percent error (MAPE), and Normalized Mean Square Error (NRMSE) are calculated.

4.1 Root Mean Square Error

The root-mean-square error is a frequently used measure of the differences between values predicted by a model or an estimator and the values observed. Here the actual software reliability value and predicted value are used to calculate the RMSE.

$$RMSE = \frac{1}{N}\sqrt{\sum_{n=0}^{N-1} |\hat{y}_n - y_n|^2} \tag{11}$$

4.2 Average Error

The average error is the difference between the actual output and predicted output which is averaged by the total number of tests performed. This is inversely proportional to the performance of the system. Less value will give better performance.

$$AE = \frac{1}{N} \sum_{n=0}^{N-1} \hat{y}_n - y_n \tag{12}$$

4.3 Mean Absolute Error

MAE is the absolute difference between the actual output and the predicted output. This metric can give a good understanding of the performance of the trained model. The mathematical equation used can be shown below.

$$MAE = \frac{1}{N} \sum_{n=0}^{N-1} |\hat{y}_n - y_n| \tag{13}$$

4.4 Mean Absolute Percent Error

This is one of the important metrics in the prediction quality evaluation technique. This can produce a value between 1 to 100. The performance of the model is inversely proportional to the value.

$$MAPE = \frac{100}{N} \sum_{n=0}^{N-1} |\hat{y}_n - y_n| \tag{14}$$

4.5 Normalized Mean Square Error

This metric produces a value between 0 to 1. This value is normalized to one for better understanding. The mathematical equation which is used to calculate the performance can be shown in the below mathematical equations.

$$NMAPE = \frac{100}{N} \sum_{n=0}^{N-1} \frac{|\hat{y}_n - y_n|}{y_n} \tag{15}$$

Table 1 shows the RMSE performance for software prediction for various input parameters. The KGMO-ANN is trained with different parameters to validate the performance of the proposed algorithm. Also, other techniques such as ABS-ANN, PSO-ANN, GA-ANN, ANN used to perform the comparative analysis. From the table, it is very clear that the proposed method reduces the RMSE.

The quality of the algorithm is inversely proportional to the RMSE value. If the RMSE value is less then the performance of the algorithm is good. Table 2 shows the performance of various metrics for software reliability prediction for different techniques. From the table, it is very clear that the proposed one is giving less error when compared to other software prediction techniques. From Table 1 and Table 2, the average of the error value is very less in the proposed method when compared to the other conventional techniques.

Figure 3. shows the performance metrics for software reliability prediction in chart format. From Fig. 3. the RMSE value for the proposed method is getting reduced for different input parameters. The RMSE for KGMO-ANN for the CBO parameter is 0.7 while RMSE of ANN is 0.93. also for other techniques, the RMSE is comparatively high.

Table 1. RMSE for software prediction for various input parameters

No of parameters	KGMO-ANN	ABS-ANN	PSO-ANN	GA-ANN	ANN
Cbo	0.89	0.88	0.85	0.90	0.93
Dit	0.88	0.89	0.95	0.85	0.94
fanIn	0.85	0.88	0.91	0.93	0.94
Lcom	0.97	0.85	0.94	0.94	0.91
Noc	0.91	0.94	0.93	0.93	0.96
numberOfAttributes	0.94	0.91	0.85	0.94	0.95
numberOfAttributesInherited	0.93	0.94	0.97	0.95	0.97
number Of Lines of Code	0.96	0.89	0.91	0.91	0.92
numberOfMethods	0.94	0.89	0.94	0.94	0.95
numberOfMethodsInherited	0.95	0.94	0.97	0.96	0.98

Table 2. Various performance metrics for software reliability prediction

	RMSE	AE	MAE	MAPE	NMSE
ANN	0.85	0.88	0.85	0.82	0.91
ABS-ANN	0.88	0.89	0.88	0.85	0.94
PSO-ANN	0.9	0.95	0.91	0.94	0.93
GA-ANN	0.92	0.85	0.93	0.94	0.93
KGMO-ANN	0.93	0.94	0.94	0.91	0.96

Also Fig. 4. shows the various performance of the proposed techniques. Here all performance evaluation metrics for software reliability are evaluated. In this, All the performance metrics are comparatively less when compared to the other techniques.

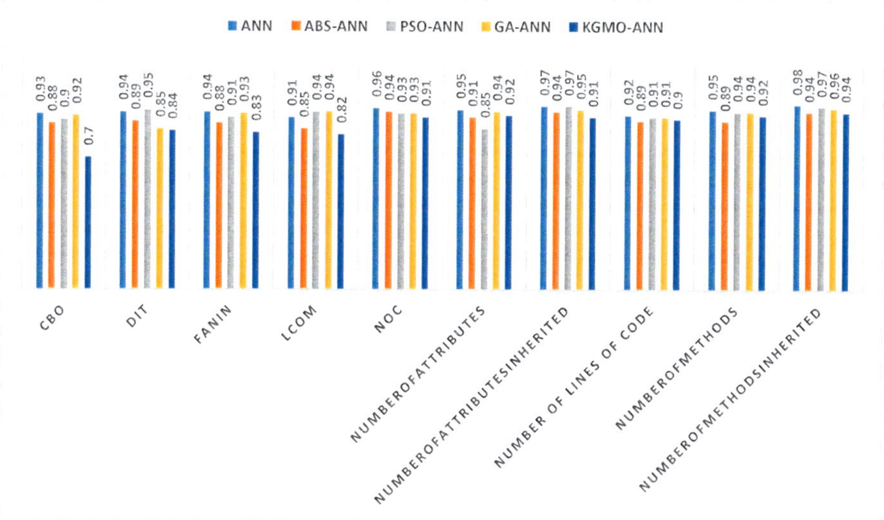

Fig. 3. RMSE for software prediction for various input parameters

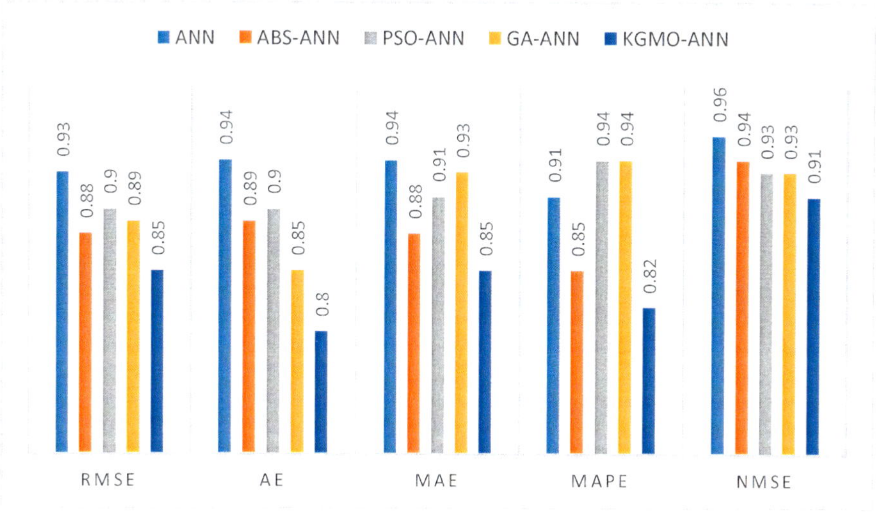

Fig. 4. Various performance metrics for software reliability prediction

5 Conclusion

In this work, software reliability prediction is done based on the various parameters using KGMO based weight-optimized neural network. Training is performed by pre-processing the data by normalization and scaling followed by the data cascading process. Various performance evaluation techniques are used to evaluate the performance of the proposed techniques. It is found that the proposed technique for software reliability prediction is giving the best performance when compared to conventional software reliability prediction techniques.

References

1. Li, Q., Pham, H.: A generalized software reliability growth model with consideration of the uncertainty of operating environments. IEEE Access **7**, 84253–84267 (2019)
2. Diwaker, C., et al.: A new model for predicting component-based software reliability using soft computing. IEEE Access **7**, 147191–147203 (2019)
3. Gheisari, M., et al.: An optimization model for software quality prediction with case study analysis using MATLAB. IEEE Access **7**, 85123–85138 (2019)
4. Arshad, A., Riaz, S., Jiao, L., Murthy, A.: Semi-supervised deep fuzzy c-mean clustering for software fault prediction. IEEE Access **6**, 25675–25685 (2018)
5. Zou, B., et al.: Reliability analysis and allocation: development of a hierarchical structure modeling platform in I&C system software life cycle. Nucl. Eng. Des. **328**, 345–352 (2018)
6. Yadav, H,B., Yadav, D.K.: Early software reliability analysis using the reliability of relevant software metrics. Int. J. Syst. Assur. Eng. Manag. **8**(4) 2097–2108 (2017)
7. Wang, J., Zhang, C.: Software reliability prediction using a deep learning model based on the RNN encoder–decoder. Reliab. Eng. Syst. Saf. **170**, 73–82 (2018)
8. Arshad, A., Riaz, S., Jiao, L., Murthy, A.: Semi-supervised deep fuzzy c-mean clustering for software fault prediction. IEEE Access **6**, 25675–25685 (2018)
9. Li, Z., Miaomiao, Y., Wang, D., Wei, H.: Using a hybrid algorithm to estimate and predicate based on software reliability model. IEEE Access **7**, 84268–84283 (2019)

The Next Step Is Nanotechnology: Application and Recent Advancement

Archit Gupta(✉) and Princy Randhawa(✉)

Department of Mechatronics Engineering at School of Automobile, Mechanical and Mechatronics Engineering, Manipal University, Jaipur, India
Archit.179403012@muj.manipal.edu

Abstract. Nanotechnology is receiving great attention in the Production industry and is entitled to be the technology of the future. With their impressively unique physical and chemical characteristics due to the large surface area and nano (10^{-9} m) size, nanomaterials are significantly enhancing the structure of the Manufacturing Industry. Some of the manifold applications reach from Sustainable Development, Pollution Sensing, Driving dynamics up to Light Weight construction, and much more.

This paper reviews the application of this technology in the field of Material Processing, Energy, Environmental Remediation, Medical, and Healthcare. Moreover, the paper analytically examines the advances and applications of nanotechnology in the Automobile Industry. Also, some insights on self-healing materials, Light Weighting, and Ultra Reflective mirror occupied in the next generation of Vehicles have been introduced.

An overview of the Future of Nano-Automotive Market Planning and the effect of COVID-19 on the Nano-Automotive Market is also included. The Compound Annual Growth Rate (CAGR) by the year 2025 and the viability of nanomaterials in future projects of multiple Automotive Companies have also been discussed.

Nanotechnology would certainly have an immense impact on not just one but practically all industries and even on the environment, but the result of this study is limited towards the fact that Nanomaterials could be the basis of a new generation of Stronger, lighter, and efficient self-healing materials.

Keywords: Nanotechnology · Automobile · Environment remediation · Nano-automotive · Applications · Self-healing material · Lightweight materials · Future technology · COVID-19 impact

1 Introduction

Nanotechnology has fetched incredible variations in the industries that are re-innovating their stance. It is made possible to manufacture products that are more competent, smart looking, tougher and robust, all because of their enormous applications (Lanzara et al. 2009).

In near future, it is anticipated that an efficient set of research tools and clinically accommodating devices would be seen. Innovative drug delivery systems, virtual imaging, and new therapies, would be the futuristic applications of nanotechnology, concluded

L. Garg et al. (Eds.): ISMS 2020, LNNS 303, pp. 171–183, 2022.
https://doi.org/10.1007/978-3-030-86223-7_16

by The National Nanotechnology Initiative (CISN - Why Nanotechnology Is Important n.d.).

Approximately 1.5 billion vehicle fleet will double from 750 million today by 2030, is estimated by the United Nations. Mainly due to swiftly growing markets in China, India, Russia, Brazil, and South Africa. Every year, roughly 50 million new cars are manufactured in the world, which directs that nanotechnology and its products have enormous opportunities in the automobile industry (Nanomaterials in the Automotive Industry n.d.).

Alloy steel is used by the automobile companies for its body and has been tremendously successful, but the industry is still looking towards how to tackle with corrosive properties of steel (Sheng et al. 2015). The alternative to steel, because of its heaviness, is to use aluminum but this replacement can happen to make it expensive. Here comes the usage of Carbon Nano Tubes, because of its strength, hardness and bending properties, better than both steel and aluminum and cheaper than two metals.

1.1 What is Nanotechnology?

The science of building nanoscale structures at the order of 10–9 m is Nanotechnology.

Changes in chemical, physical, and biological properties are observed, when the proportions of a material is concentrated to nanoscale which vary significantly from the characteristics of the atoms and molecules of the original size.

These materials showcase modified characteristics like Surface area to volume ratio, molecular confinement, and abridged imperfection (Tseng 2013).

At the nanoscale, increase in surface area is observed, leading to more atoms being exposed thereby making the properties of the bulk material less dominant than surface properties. These upgraded characteristics assist in confrontation against, corrosion, oxidation, and extreme temperatures (Recent Advances of Nanotechnology in Transportation n.d.; Sheng et al. 2015).

2 Future Applications of Nanotechnology

Nanotechnology has been assistive to not only improve but even revolutionize many technologies and sectors of industry including information technology, medicine, automation, energy, homeland security, and environmental science only to name a few (Benefits and Applications|Nano n.d.) (Fig. 1).

2.1 Industry and Material Processing

Nanotechnology has backed up major advances in the field of computing and electronics, heading to smaller, faster, and lighter systems that could bring about high amounts of data.

Similarly, nanotechnology has played a significant role in the Automotive Industry as can be witnessed with the Improvement in Cars, Marines, and Aircraft over the years (Benefits and Applications|Nano n.d.).

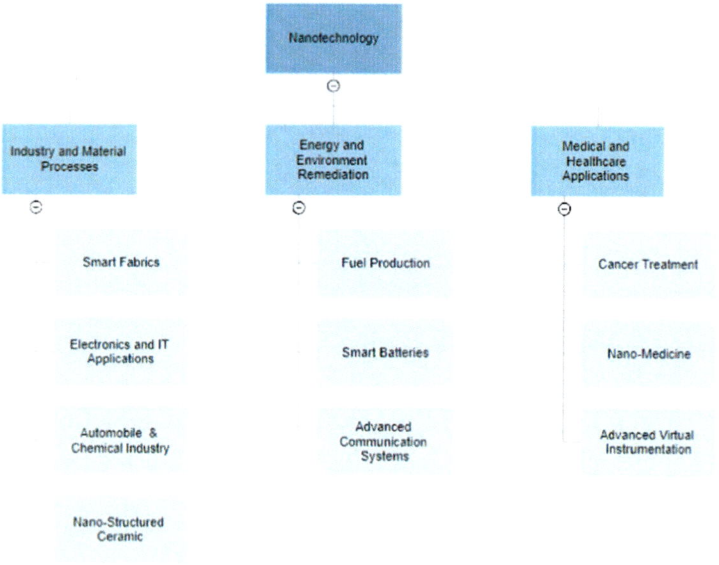

Fig. 1. Future application chart of nanotechnology

- Nanomaterials are a foundation to an era of washable, efficient "Smart Fabrics" engaged with capabilities for monitoring health, harvesting energy through movement, and capturing solar energy all with the assistance of nano sensors and electronics
- Industries where Smart Fabrics have already been put to use include Stain repellent and wrinkle-resistant threads, Bodywarmers, Nano socks treated with silver nanoparticles, etc. (Application of Nanotechnology in Textile Industry, Nanotechnology Application in Textiles-Fibre2Fashion n.d.)
- Nano-engineered materials as automotive products include Self-Healing Paint Coating, Light weight-high strength materials, Ultra-Reflecting Mirrors and Glasses, and many other applications (Mathew et al. 2019).
- Magnetic random-access memory (MRAM) is inculcated with nano-scale magnetic tunnel intersections that efficiently back up data when system goes through a shutdown and enable features like resume-play.

2.2 Energy and Environment Remediation

Alternative Energy approaches are crucial to meet the world's escalating energy demands. Research is been done to find clean, inexpensive, and sustainable energy sources.

Traditional methods are being transformed in modern ways of energy harvesting with help of nanotechnology (Fig. 2).

- Raw petroleum materials through the integration of catalysts are being used to improve the efficiency of fuel production. Reduced fuel consumption in automobiles is being commissioned due to higher-efficiency combustion and reduced friction.

Fig. 2. Nanoparticles fused to make new Flexible Solar panel Cells (Image courtesy of Nanosys)

- Fast-charging, lighter in weight, efficient, higher power density, and could hold electrical charge longer batteries are being developed with the implementation of Nanotechnology.
- In-expensive detection and treatment of contaminations in water could assistance in the need for approachable problem, hygienic drinking water.
- "Paper towel" which hold 20 times its weight in oil is being woven with small wires of Potassium manganese oxide (Fig. 3).

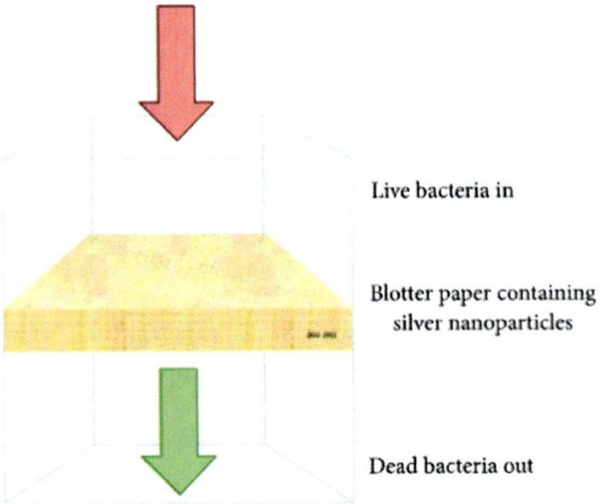

Fig. 3. Silver nanoparticles being used for disinfection process. (Saleh 2017)

2.3 Medical and Health Care Applications

Disease anticipation, recognition, and treatment using nanotechnology are called Nanomedicine. Producing target specific medicine at the nanoscale (Fig. 4).

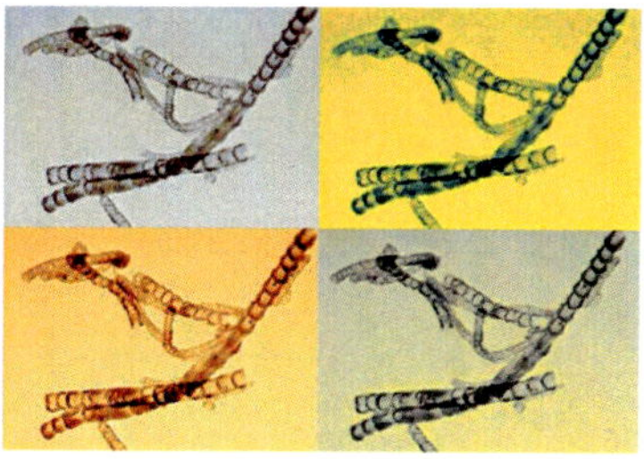

Fig. 4. Treatment of cancer with nitrogen-doped carbon nanotubes (Source: Wake Forest and the National Cancer Institute)

- The way of treating cancer is getting revolutionized, drastically minimizing the unhealthy effects of Chemotherapy. Researchers are employed upon various techniques where nanocontainers could directly deliver medication to cancer cells and reduce the danger of mutilation to the fit tissue.
- Technology beyond the usage of needles is being introduced. Researchers are exploring ways that nanoparticles could improve vaccines, including vaccine delivery (Nanotechnology|Nanomedicine|Nanotechnology n.d.).
- Better diagnosis, efficient treatments, and successful therapeutic rates because of improved imaging and analytical tools blessed by nanotechnology.
- Graphene nanoribbons are being researched upon for the healing of Spinal Cord Injuries.

3 Applications of Nanotechnology in Automobiles

By the year 2024, the worldwide nanotech market is anticipated to surpass a net value of almost $124 billion USD (Analysis 2020). As the automobile sector is one of the world's largest economic sectors by revenue utmost research and expansions based on nanotechnology are done here.

We are going to discuss 3 future applications on which the Automobile Industry is employed (Fig. 5).

3.1 Self-healing Material

These are a branch of Smart Materials that showcase the ability to repair themselves from structural damage caused by mechanical or external force overtime.

Terzo Millennio is the world's First self-healing sports car by Lamborghini and in collaboration with Massachusetts Institute of Technology (MIT) researchers is being drafted in Boston.

The sports car is integrated with its private health monitoring system which could detect cracks and damages internally after the accidents (Lamborghini Creates World's First "self-Healing" Sports Car n.d.).

When the car senses damage to the exterior, the crack heals by the heat produced by the microchannels.

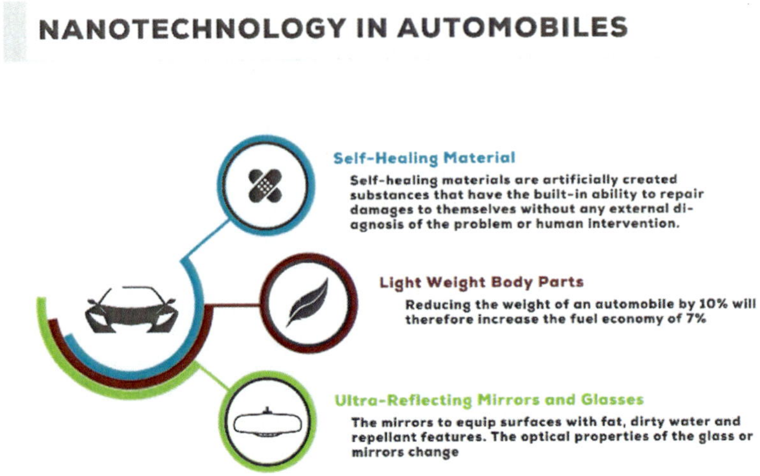

Fig. 5. Contribution of Nanotechnology in Automobiles.

As carbon fiber is applied extensively all over the car, the weight of the vehicle is also subsequently decreased, says Lamborghini (Lamborghini Terzo Millennio Can Store Electric Energy and Heal Itself n.d.) (Fig. 6)

Fig. 6. Lamborghini Terzo Millennio.

How does Self-healing Work?
Materials that exhibit characteristics to heal themselves and recover after mutilation and/or without outside intervention are called Self-healing materials. The process takes place in two types: Covalent and Non-Covalent (Menikheim and Lavik 2020). Intrinsic, vascular, and capsule-based are the three Self-healing mechanisms (Fig. 7).

The Most Effective Approach: Using Nano Containers
Here an active material is kept in a permanent trapped state, dodging contact between the active and the passive section, leading to leakage. The nanocontainers are uniformly

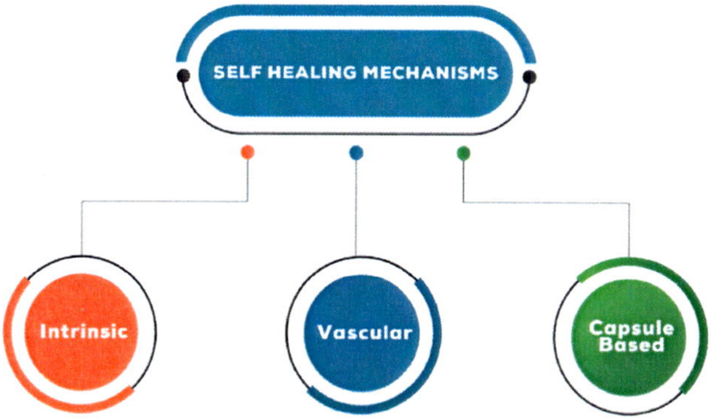

Fig. 7. Ways of Self Healing.

spread in the passive matrix. When the matrix experiences any external changes or disturbances, nanocontainers respond to this and release the active material (Nanotechnology Material, Heal Thyself n.d.; Shchukin and Möhwald 2007).

Dimensions of these nanocontainers must be less than 300–400 nm; deep cavities are formed, in nanocontainers of the bigger size, which could subsequently harm the integrity of the matrix (Shchukin and Möhwald 2007) (Fig. 8).

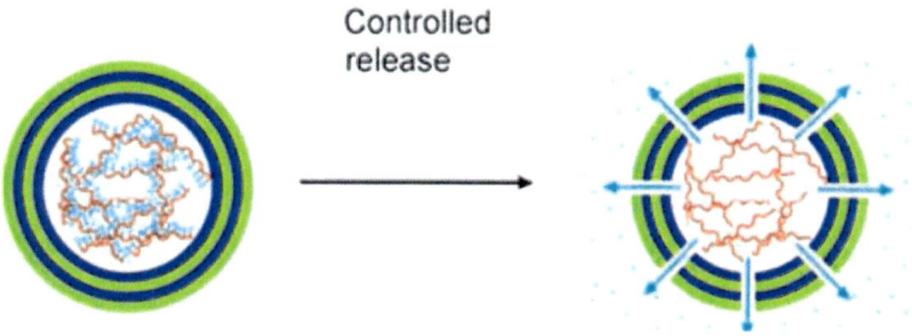

Fig. 8. Source: (Nanotechnology Material, Heal Thyself n.d.)

Self-healing Paint Coatings (Nano Passivation)

Pre-treatment, electrodeposition, a sealer, a primer, and the topcoats are the 5 modern steps the automotive coating.

Conventional paints came with its own problems, low paint transfer, minor flame resistance characteristics, fewer surface finish, water, and dust particles being adhesive (Mathew et al. 2019).

The application of the zinc layer is initiated here, constituting of SiO_2 nanoparticles (Fig. 9). Nano passivation is the name of this self-healing mechanism.

Mercedes Benz is being employed with these types of nano paints.

Binders (in the orange) can be seen in Fig. 10(a) and the cross-linker agent (in the red) of historical paints. Better elasticity is shown in Fig. 10(b) (yellow) composed of nanoparticles) in nano paints having great strength (blue).

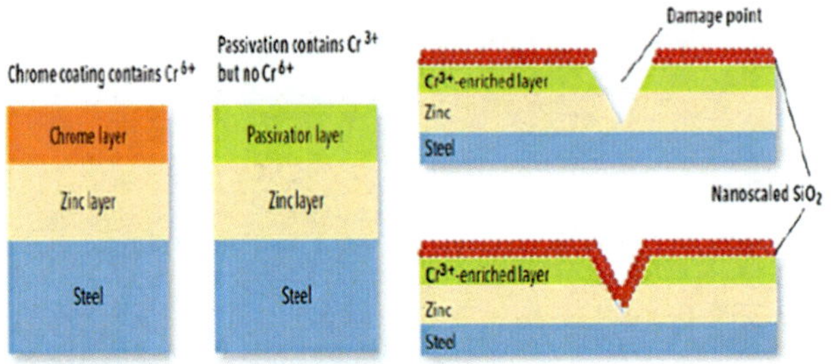

Fig. 9. Source: (Industry n.d.)

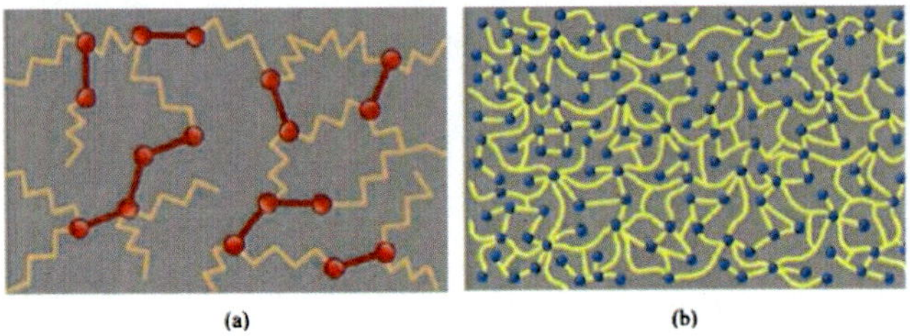

Fig. 10. Scratch-resistant characteristics are induced to the paint due to the closely packed nanoparticles. Source: (Industry n.d.)

This also has an application in Marine Transportation. Punitive seawater and atmosphere lead to corrosion on the ship, which affects the efficiency of the cargo being transported.

3.2 Light Weighting: A Weight Loss Pill for the Automotive Industry

Automobile Industries' most unsettled debate of all time is the weight reduction of vehicles. Maximizing fuel efficiency is done by cutting the extra weight on the vehicle, leading to reduced CO_2 emissions, and lower manufacturing cost. Approximately 7% of the fuel economy increases by a reduction of 10% in the weight of the automobile (Recent Advances of Nanotechnology in Transportation n.d.).

A Leapfrog from BSVI from BS IV – Necessary Pollution Control

- The Indian automotive Industry has taken a leapfrog from BS IV to BS VI, skipping BSV. BS VI sets down a low 10 ppm of Sulphur. Taking pollution into consideration caused by vehicles of BS V, weight reduction was necessary.
- The BS6 emission norms have been reduced to 60 mg/km of NO_2 (nitrogen oxides), while the BS4 emission norms allowed a motor vehicle to emit not more than 80 mg/km of NO_2

Nanotechnology application into the automotive industry primes to lighter automobile bodies without negotiating with rigorousness and crashworthiness, leading to low material and indirectly less fuel ingestion.

Carbon Nanotubes (CNTs) are being called the ideal substitute for steel in automobile Industry, these provide high strength (150× Steel) and decreased weight (1/5× Steel) (Coelho et al. 2012).

- Using nanocomposites and hybrid technology, 900 kg of steel could be condensed by up to 300 kg.
- The weight reduction of around 25% over filled plastics and 80% over steel could be attained with the aid of nanocomposites (Coelho et al. 2012) (Fig. 11).

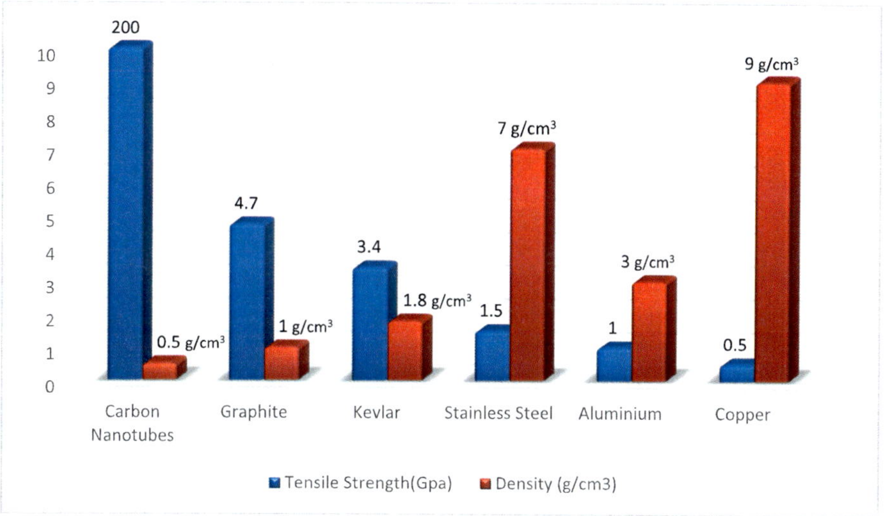

Fig. 11. Comparison of Nanomaterials (Carbon Nanotubes) with other materials in terms of Strength and Density.

Future Automotive Market Planning

- The Volkswagen Nano-spyder is a fully recyclable project. It would comprise billions of nanomachines that could arrange the vehicles and then be fragmented apart, ready for reuse in a new vehicle (Applications of Nanotechnology in Car Manufacturing|ManufacturingTomorrow n.d.; Guo et al. 2014; Regt et al. 1995) (Fig. 12).
- Tesla is working on sports cars, using lightweight carbon fiber to install tanks and sink. The Body panels are being made in form of a sandwich, glass and polypropylene are between the two carbon layers.

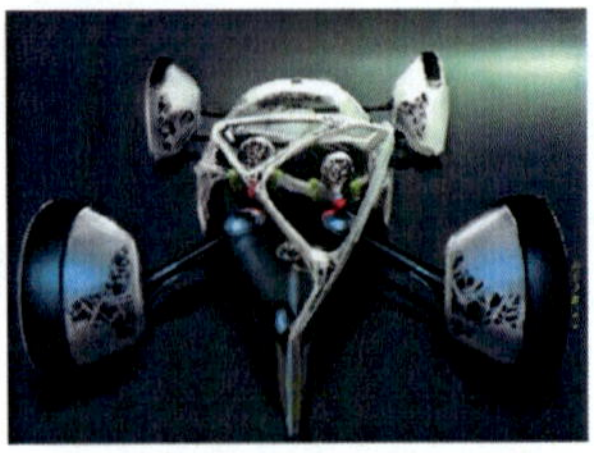

Fig. 12. Spyder Car by Volkswagen (Nanospyder et al. (2020))

- Around 8 kg of weight reduction by Volkswagen will be done using long-fiber-reinforced thermoplastic (LFRT), eliminating the metal and processing steps.
- Having safety as a priority, Ford Motor Company has set its goal to diminish weight by 340 kg. The company approaches to advance the fuel economy by 40%.

3.3 Ultra-reflecting Mirrors and Glasses

During the day and late evening driving time, the distress (glare of the sunlight and approaching vehicles) while driving could be expressively reduced by the help of ultra-thin reflective mirrors (Industry n.d.).

Dirt and water repellant mirror surface features are offered by this ultra-thin layer of aluminum oxide. The chemical vapor deposition (CVD) technique is used to form these nano-thick layers called Hydrophobic layers (Mohseni and Ramezanzadeh 2012). Figure 13a shows a conventional mirror while Fig. 13b shows a modern hydrophobic layer-based mirror (Fig. 14).

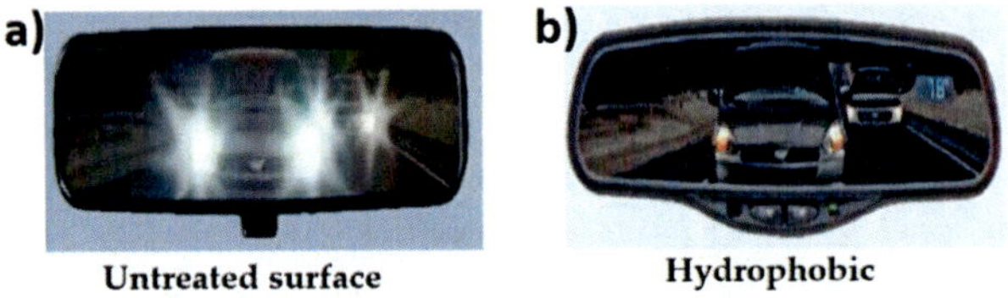

Fig. 13. (a) Historical mirror and (b) current antiglare mirror (hydrophobic) (Mohseni and Ramezanzadeh 2012)

Fig. 14. Titania nano film on the side view mirror. Source: (Athar 2010)

4 COVID-19 Impact on Nano-automotive Market

At a CAGR (Compound Annual Growth Rate) of 7.3%, from USD 69.7 billion in 2020 to USD 99.3 billion by 2025, could be estimated growth of The Automotive Material Market, post COVID-19. The expanding interest for lightweight materials is driven by the rising utilization of electric vehicles and rigorous emission norms. CO_2 reduction targets have been set by several OEMs for the coming years (Fig. 15).

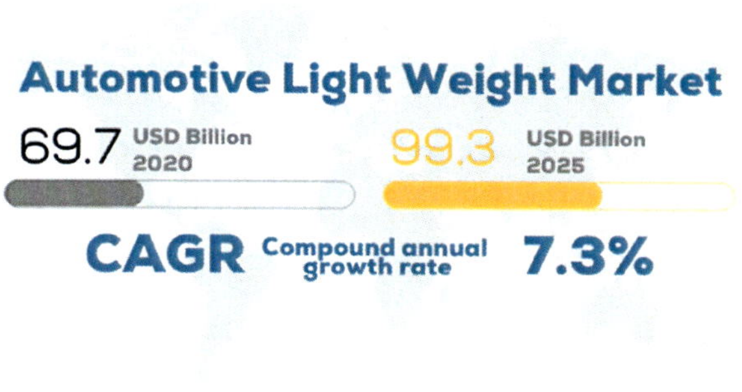

Fig. 15. Compound Annual Growth Rate of Automotive Light Weight Market by the year 2025.

5 Conclusion

This review paper has highlighted the extraordinary properties of Nanomaterials which have countless applications in the field of Automotive, Energy, Environment Remediation, Medical, and Health Care.

In the 21st century of transforming era. Nanotechnology is an answer to enhancing and revolutionizing the quality of living beings. The changes brought by the technology are tremendously affecting the future of the Industries. It keeps evolving at a fast pace and is the driver of the new concept of digitalization. Nanotechnology is the key that has made transportation durable, stronger, efficient, and smart.

With the scarcity of natural resources (steel and aluminum) and the non-biodegradable nature of plastic, making it an environmental hazard; Carbon Nanotubes (CNTs) are nothing but the best replacement. Although to upsurge the feasibility of CNTs, the purification process of the same needs to improve. The pandemic of COVID-19 has caused a major upheaval in the supply chain of the system due to the halt in manufacturing facilities across the globe. It is anticipated that the market is going to have serious impacts since the development of the market of lightweight material is in a straight line with production.

Nanotechnology is soon going to be the new normal. Nanostructures, nanomaterials, and nano engines will be controlling the industries and lead to the betterment of life. Though the exposure and toxicity of nanomaterials are a major alarm and require immediate consideration to ensure a clean and fit environment.

Since nanomaterials are new to the environment, therefore it is essential to understand their latent effects in and across air, soil, and water. It is needed that in-depth research work on nanotechnology and its environmental interaction be carried out, along with imminent international cooperation and collaboration in checking the environmental challenges faced. The development of an effective impact assessment system could better assure us of a safe and sustainable application of nanotechnology in the multiple industries.

Judging by the technology taking leaps, the day is not far off when nanotechnology will rightfully fit in the Future Industry cycle.

References

Analysis, M.: Biomaterials and Medical Applications Nanotechnology 2020 Market Analysis, vol. 3, no. 1, pp. 2019–2020 (2020).

Application of Nanotechnology in Textile Industry, Nanotechnology Application in Textiles-Fibre2Fashion (n.d.). https://www.fibre2fashion.com/industry-article/7135/application-of-nanotechnology-in-textile-industry. Accessed 13 Sep 2020

Applications of Nanotechnology in Car Manufacturing|ManufacturingTomorrow (n.d.). https://www.manufacturingtomorrow.com/article/2018/09/applications-of-nanotechnology-in-car-manufacturing/12188. Accessed 15 Sep 2020

Athar, T.: Metal oxide nanopowder, 1st edn. In: Emerging Nanotechnologies for Manufacturing. Elsevier Inc. (2010). https://doi.org/10.1016/B978-0-8155-1583-8.00014-4

Benefits and Applications|Nano (n.d.). https://www.nano.gov/you/nanotechnology-benefits. Accessed 13 Sept 2020

CISN-Why Nanotechnology Is Important (n.d.). https://cisncancer.org/research/new_treatments/nanotechnology/importance.html. Accessed 16 Sept 2020

Coelho, M.C., Torrão, G., Emami, N., Grácio, J.: Nanotechnology in automotive industry: research strategy and trends for the future-small objects, big impacts. J. Nanosci. Nanotechnol. **12**(8), 6621–6630 (2012). https://doi.org/10.1166/jnn.2012.4573

Guo, D., Xie, G., Luo, J.: Mechanical properties of nanoparticles: basics and applications. J. Phys d: Appl. Phys. **47**(1), 013001 (2014). https://doi.org/10.1088/0022-3727/47/1/013001

Automobiles Industry: Nanotechnologies in Automobiles (n.d.)

Lamborghini creates world's first self-healing sports car (n.d.). https://www.telegraph.co.uk/news/2017/11/11/lamborghini-creates-self-healing-sports-car/. Accessed 14 Sept 2020

Lamborghini Terzo Millennio Can Store Electric Energy and Heal Itself (n.d.). http://compositesmanufacturingmagazine.com/2017/11/lamborghini-terzo-millennio-can-store-electric-energy-heal/. Accessed 14 Sept 2020

Lanzara, G., Yoon, Y., Liu, H., Peng, S., Lee, W.I.: Carbon nanotube reservoirs for self-healing materials. Nanotechnology **20**(33), 335704 (2009). https://doi.org/10.1088/0957-4484/20/33/335704

Mathew, J., Joy, J., George, S.C.: Potential applications of nanotechnology in transportation: a review. J. King Saud Univ. Sci. **31**(4), 586–594 (2019). https://doi.org/10.1016/j.jksus.2018.03.015

Menikheim, S.D., Lavik, E.B.: Self-healing biomaterials: the next generation is nano. Wiley Interdisc. Rev. Nanomed. Nanobiotechnol. **12**(6) (2020). https://doi.org/10.1002/wnan.1641

Mohseni, M., Ramezanzadeh, B.: The role of nanotechnology in automotive industries, pp. 3–54 (2012)

Nano engineered materials in automotive products include high power|Course Hero (n.d.). https://www.coursehero.com/file/p6mkgd8/Nano-engineered-materials-in-automotive-products-include-high-power/. Accessed 13 Sept 2020

Nanomaterials in the Automotive Industry (n.d.). https://www.azonano.com/article.aspx?ArticleID=4936. Accessed 16 Sept 2020

Nanotechnology | Nanomedicine | Nanotechnology (n.d.). https://www.scribd.com/document/396966216/STS-40-Paper. Accessed 13 Sept 2020

Nanotechnology material, heal thyself (n.d.). https://www.nanowerk.com/spotlight/spotid=2067.php. Accessed 14 Sept 2020

Recent Advances of Nanotechnology in Transportation (n.d.). https://www.azonano.com/article.aspx?ArticleID=4826. Accessed 13 Sept 2020

De Regt, J.M., van Der Mullen, J.A.M., Schram, D.C., van Dijk, J.: Components of continuum radiation in an inductively coupled plasma. J. Phys. d: Appl. Phys. **28**(1), 40–46 (1995). https://doi.org/10.1088/0022-3727

Saleh, T.A.: An overview of nanomaterials for water technology. Adv. Nanomater. Water Eng. Treat. Hydraulics **2016**, 1–12 (2017). https://doi.org/10.4018/978-1-5225-2136-5.ch001

Shchukin, D.G., Möhwald, H.: Self-repairing coatings containing active nanoreservoirs. Small **3**(6), 926–943 (2007). https://doi.org/10.1002/smll.200700064

Sheng, Y.R., Yang, Q.Y., Zhang, Z.G.: A review on applications of seismic-fluid geochemistry. Bull. Mineral. Petrol. Geochem. **34**(4), 837–842 (2015). https://doi.org/10.3969/j.issn.1007-2802.2015.04.019

Tseng, F.: Advanced materials manufacturing and characterization Nano/Micro fluidic systems for circulating Tumor cells (CTCs) rapid detection and diagnosis. Adv. Mater. Manuf. Charact. **3**(1), 4–7 (2013)

Volkswagen Nanospyder-Car Body Design (n.d.). https://www.carbodydesign.com/archive/2006/11/01-volkswagen-nanospyder. Accessed 15 Sept 2020

Check for updates

Employee Motivation and Its Impact on Organizational Performance with Mediating Role of Leadership Approach: A Study of Select Hospitality Organizations in Uttarakhand (India)

Bhawna Chahar(✉)

Manipal University Jaipur, Jaipur, Rajasthan, India

Abstract. The present research aims at examining mediating role of leadership approach in the relationship between employee motivation and organizational performance of hospitality organization in Uttarakhand state (India). A survey using propulsive sampling of 217 employees associated with different hospitality organizations located in Uttarakhand state (India) using questionnaire as a primary data collection instrument was carried out. The research focuses on three specific variables for analysis. Organizational performance was taken as dependent variable, factors of employee motivation were taken as independent variable and leadership approach was taken as mediating variable. Four different hypotheses were proposed and evaluated using descriptive statistics and regression analysis, respectively. The research findings indicate employee motivation have positive relationship with organizational performance (β = .602, t = 8.089 and R^2 = .233). Similarly, it was found that employee motivation has positive influence on leadership (β = .757, t = 12.715 and R^2 = 429) and Leadership practices in the organization have positive influence on organizational performance (β = .668, t = 11.559 and R^2 = .383). It is also found that leadership approach has mediating effect between employee motivation and organizational performance. Study offers some meaningful suggestion to improve employee motivation and enhance organizational performance. Some of the managerial implication, limitations and scope of future research are also presented in the study.

Keywords: Employee motivation · Organizational performance · Leadership approach

1 Introduction

Business environment is changing fast in across the globe and most of the organizations realize about the employee motivation and how leadership and management can motivate their employees to give a better performance in their job and gain competitive edge. Hospitality sector is a service sector, and its performance can be determined by improving the factors like management, availability and optimization of finance and employees. Out of

L. Garg et al. (Eds.): ISMS 2020, LNNS 303, pp. 184–201, 2022.
https://doi.org/10.1007/978-3-030-86223-7_17

various components, employee is critical factor for improving performance of hospitality organizations. Emphasizing on employee performance will improve the performance of the organization. If employee performs well in their jobs, the efficiency of the company will improve indirectly (Campbell 1990). Employees typically perform better when they have a drive in their work. The companies will therefore strive hard to provide better-working conditions, health arrangements, and provide the employee with opportunities to recognize their skills and improve their motivation. The global business climate will thrive if rapidly evolving and the companies adaptable to the transition. Organizations need to construct plans to overcome the competitive rivalry, and the one that can succeed will be able to maintain themselves longer than others. Employee motivation helps organizations to increase employee job satisfaction, reduce turnover and thus improve organizational performance (Singh and Tiwari 2011). The HRM practices and policies have to be directed and aligned with organizational strategies and understanding expectations of the employees. This research attempted to explore employee motivation, and how leadership affect relationship between motivation and organizational performance.

Hospitality sector organization is the service organization where human resource play an important role in delivery of services. As such, keeping the employees motivated all the time is important for hotel managers (Çetin 2013). Employee motivation also plays a key role in the management of hotels (Wasike and Ndivo 2015). Ross (2005) stated that because of hotel industry's labor-intensive nature, workers work in stressful circumstances, they work for long, irregular hours and encounter uncomfortable situations during the customer service delivery phase. A motivated and happy worker is a prime concern in the provision of hotel services (Kingir and Mesci 2010). Hence, the workers should be kept happy and motivated, as the workers satisfaction or dissatisfaction may affect the customer satisfaction or dissatisfaction with the services rendered (Çetin 2013).

Most researchers believe that leadership style is closely related to success whereas other scholars assume that employee behavior is more related to an organization's overall efficiency and performance level. Undisputable, various academicians have different ideas about organizational behavior and leadership style and varying expectations. The question remains, which leadership style is better for optimizing success of organizations and employees? What difference does leadership styles bring to success level? Are there positive or negative associations between leadership and organizational behavior, and employee performance? These constitute the major problems that triggered and motivated this study. The present study is undertaken to study the employee motivation and its relationship with organizational performance with moderating role of leadership of hospitality organization in Uttarakhand (India). Hospitality industry is chosen as it is growing at a fast pace. The state of Uttarakhand is primarily known as religious and spiritual tourism destination in the country. Human resource determines the success or failure of an organization. Leadership in an organization keeps its employee motivated and help in improving organization's performance and productivity. Uttarakhand is one of the most preferred tourist destination state in India, but it also faces a lot of problems in hospitality industry, mostly because of its geography as well as trained and motivated manpower. Finding the means of employee motivation and how leadership of the organization helps in motivating employees and boost organizational performance is of relevance to the industry. All these reasons justify the relevance of undertaking this study for better decision making. Therefore, the findings of this study will be valuable for hospitality organization in framing suitable strategies to motivate employees for better organization outcomes in this emerging economy.

2 Review of Related Literature and Development of Hypothesis

Employee Motivation and Its Relationship with Organizational Performance
Studies have shown that the understanding of workers about the organization's goals and norms is positively linked to employee motivation. This cycle also contributes to organizational development, given that empowerment helps in achieving high motivation rates. Bartol and Martin (1998) concludes that motivation strengthens actions and activates a continuing propensity. Motivation is an intrinsic desire to accomplish a certain goal that starts with a physiological or psychological need and activates an objective-set output. Employee performance is dependent on variables like motivation, work satisfaction, appraisals, training, and growth, etc., but employee motivation is primarily focused on this research since it affects the organizational performance to a certain extent. Kalimullah (2010) indicated that motivated employee's interests are aligned with the interests of the organization thus focusing its efforts on that direction. Moreover, employees of these organizations are more effective because they are always searching methods that can better their jobs. The employees need to be motivated so that they meet their full potential at work under demanding situations. Numerous research have concluded that bonus contributes to work satisfaction that influences directive and employees' performance in a positive way. To motivate employees for improving their performance, most of the businesses use different types of incentives. Depending on the work performed by each employee and its importance to the organization, employee performance and allowances, managers need to establish pay systems and use it as a motivator. Employees can also be motivated by the inspirational leadership as the leader will gain the confidence and support of the employees to achieve these goals. Nonetheless, employees should be empowered so that their trust leaders and complete the work properly for the organization (Baldoni 2005). Mary (1996) describes to what extent organizational effectiveness utilize its resources without putting burden on its members and fulfill its goals. The goal model defines organizational success in terms of the degree to which an organization meet its goals (Zammuto 1982), while the concept of program resource defines it in terms of the organization's negotiating power and its ability to influence the environment as it acquires valuable resources (Yuchtman 1987). In a survey of 151 employees from the IT sector, Kieu (2010) found that transformational leadership is a stronger analyst for organizational success, employee happiness, and loyalty by building employee commitment, confidence, trust, and respect for leaders. Intellectual stimulation and inspiring motivation have been found to enhance efficiency of the company (Hancott 2005). Based on the studies on leadership style and organizational performance, following hypothesis was proposed.

Hypothesis 1: Employee Motivation Have Positive Relationship with Organizational Performance
Employee Motivation and Its Relationship with Leadership

Leadership helps managers and administrators to impact the behavior of employee's in an organization. The successful leadership results in a greater number of motivated workers. Author such as (Koçak and Özüdoğru 2012) and Tunçer 2011 explains leadership as the process of manipulating others' actions with the intention of attaining certain goals and inspiring others during this phase. Koçak and Özüdoğru (2012) conducted surveys of 340 public and private hospital employees and found a strong link between leadership and motivation of the employees. The employee motivations that lead to the manager's demonstrated leadership characteristics and behaviors are growing. Zafar and Raja (2015) believe inspiration as central feature of leadership. Leadership is an opportunity to motivate employees so that they contribute towards the success of the organization. Asrar-ul-Haq and Kuchinke (2016) stated that employee motivation is extremely critical in public domain because the success of governments and their administrations affects the society even more than a private organization. Touré-Tillery and Fishbach (2014) differentiated, in terms of calculating motivation, between result-oriented motivation to complete a task and process-oriented motivation that implies systematic elements of goal pursuit. Touré-Tillery and Fishbach (2012) compared intrinsic and extrinsic behavioral motives and promoted context-building, for example by experimental design, to explain and distinguish one of those dimensions. Touré-Tillery and Fishbach and Choi (2012) assessed motivation in the form of successive behavior by adhering to the legal, social, and correctness parameters while discussing the result-focused dimension of motivation. It was also suggested that mean-oriented motivation is U-shaped, as it is understood that beginning and finishing behaviors are more rational for self-interpretation. Global and Chowdhury (2014) found that employees were more motivated with transformational style of leadership than transactional or laissez-faire styles. Riedle (2015) found a close positive relationship between transformational leadership style and intrinsic motivation, and relationship between transactional leadership style and extrinsic motivation. Based on studies on leadership style and organizational performance, following hypothesis was proposed:

Hypothesis 2: Employee Motivation Have Positive Influence on Leadership Leadership and Its Impact on Organizational Performance

Leadership is one of the primary factors that drive the firm towards success and enhancement. Leaders takes key - decisions on acquiring, creating, and implementing the resources and converting them into useful services & products and giving value to organizational stakeholders. Leaders determine beliefs, community, tolerance for change and motivation for employees. Intellectual stimulation and in spiritual encouragement were found to increase the efficiency of the organization (Hancout 2005). Researchers like (Meyer and Heppard 2000; McGrath and MacMillan 2000; (Judge et al. 2002b; Yukl 2002; Judge and Piccolo 2004; Purcell et al. 2004; Keller 2006) explored the strategic role of leadership, and investigated the usage of leadership paradigms and leadership behaviors to enhance organizational performance. In organizational view, it is necessary for the leader to internalize disruptive conduct to meet the competitive expectations of stakeholder in a transparent and ethical manner, since the survival of the irganization may depend on it. Intangible assets such as leadership styles, culture, knowledge, competency, and motivation combine people and processes with organizational performance (Purcell et al. 2004), p.1). Considering the organization, to meet the stakeholder's competitive standards in a manner that is clear and ethical, it is important for the leader to internalize transformational behavior, as the organization's survival may depend upon it. Leader determines the pattern of job behavior, mission activity and ethnicity of the subordinates with his organizing power. In their research, Garcia-Morales et al. (2012) found a positive relationship between leadership and success in an organization. Danişman (2015) explored how leadership affects organizational performance. A meta-analysis of 598 research studies found impact of leadership on organizational performance on a medium level. Based on the studies on leadership practices and organizational performance, following hypothesis was proposed:

Hypothesis 3: Leadership Practices in the Organization Have Positive Influence on Organizational Performance.

Hypothesis 4: Leadership Practices in the Organization Mediates the Relationship Between Employee Motivation and Organizational Performance.

3 Objectives and Research Methodology

a. To assess the relationship between employee motivation and organization leadership practices
b. To assess relationship between leadership practices and organizational performance
c. To explore the factors of employee motivation and its impact on organizational performance and assess the moderating role of leadership practices in the relationship between employee motivation and organizational performance

To conclude the existing literature review related to study variables and hypothesize the relationship among social media marketing, brand loyalty and customer experience, a conceptual framework was designed to show the proposed relationships among the variables (Fig. 1).

Descriptive research design was used to accomplish stated objectives of present research work. It involved in-depth interviews (one-on-one interviews) to fully understand the motivational practices, preferred leadership practices and perceived organizational outcome at the select hospitality organizations in Uttarakhand state (India). This

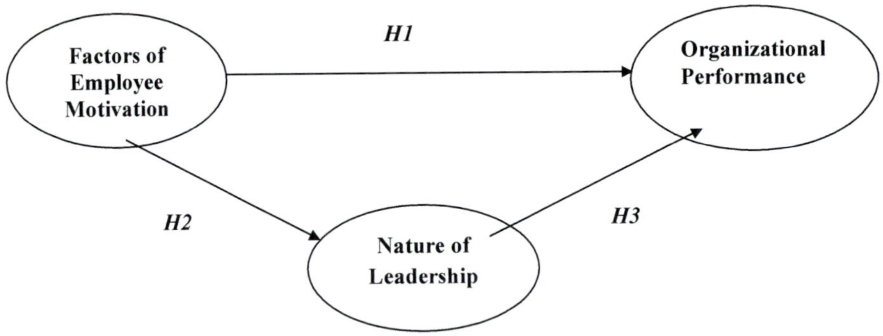

Fig. 1. Conceptual framework

approach has some promising directions for employee motivation and performance and how leadership practices help in strengthening the relationship between these two variables. Population for the present study includes employee of all the level (Management and staff level). Convenient and justified sampling techniques was used to collect the information from the employees. Since Uttarakhand state (India) is primarily known for religious and tourism destination in the country as well as in the world. Uttarakhand is a holiday destination for tourists because of its historical, cultural, and natural heritage. A lot of hospitality sector organization are established across the state irrespective of its geographical constraint. As population size is large, to keep the coverage of population, researcher has tried to adapt purposive sampling, a non-probability sampling technique where researcher has chosen participants who had specific expertise or insight regarding the subject. Sample size of 217 was chosen conveniently and justifiably that serves our purpose. Researcher collected data from different destinations in Uttarakhand (India) like Dehradun, Rishikesh, Haridwar, Nainital, Srinagar, Uttarkashi, Badrinath and Kedarnath which are well known tourist destination. Both primary and secondary data was collected to achieve stated objectives. Different sources like books, magazines, research journals, internet source, and newspapers were referred for collecting secondary data. A well-structured questionnaire covering different dimensions of employee's motivational practices, leadership practices and organizational outcomes were designed. Construct for employee motivation and measurement variable were developed and adopted from the research work of Kingir and Mesci (2010), Karatepe and Uludağ (2007) for this study. Important construct for measuring employee motivation were rated on the following variables like nature of job, job recognition, rewards and incentives, career opportunity, recognition for the work, work autonomy and individual organization fit, etc. The leadership practices were measured primarily on four dimensions i.e., direction, coaching, facilitating, and delegating and measurement variable were developed by adopting the previous researchers like Jong and Hartog (2007) and questions related to organizational performances were incorporated after reviewing Ali et al. (2010). After identifying the construct for the study, measurement variable was constructed based on the previous studies and research conducted by different authors. The questionnaire was piloted on a sample of 250 respondents associated with hospitality organization located in Uttarakhand (India). Almost 225 responses were received, after editing 217 responses of the employees were found fit and were taken for the analysis of the data. To measure the internal consistency of the responses, reliability statistics was calculated using SPSS software and found to be 0.652 to 0.953. The reliability statistics indicates that data is reliable. Mean, Standard Deviation, ANOVA test and regression analysis were carried out to analyze factors of employee motivation, leadership practices, and its relationship with organizational performance and whether leadership moderates the relationship between these two variables. First outcome presented in the table reflect the demographic characteristics of respondents.

The sample profile of respondents as presented in Table 1 denotes demographic characteristics of respondents. It is observed that majority of respondents falls here in the age group of 26–35 years with 47.5% respondents. 15.2% employee are between 36–45 years of age. 6.9% respondents belong in the age group 46–55 years. 10.6% respondents indicated that they are in the age group of above 55 years. Only 19.8%

Table 1. Sample profile

	Categories	*Count*	*Percentage*
		217	100
Age wise classification	Less than 25 years	43	19.8
	26–35 years	103	47.5
	36–45 years	33	15.2
	46–55 years	15	6.9
	Above 55 years	23	10.6
Gender	Male	171	78.8
	Female	46	21.2
Marital status	Single	84	38.7
	Married	133	61.3
Education Level wise classification	Graduate	163	75.1
	Postgraduate	26	12.0
	Professional	27	12.4
	Others	1	.5
Current position	Manager	60	27.6
	Supervisor	133	61.3
	Executive	24	11.1
Year of experience in hospitality organization	Less than 5 years	48	22.1
	6–10 years	99	45.6
	11–15 years	65	30.0
	16–20 years	5	2.3
Category of hotel	Luxury hotels	58	26.7
	Mid-scale hotels	62	28.6
	Budget hotels	93	42.9
	Heritage hotels	4	1.8

respondents found to be below 25 years of age. Majority of the employees are male (78.8%) and married categories (61.3%) respondents in the sample. It is observed that sample is a combination of educated class of employee as 75.1% employees are educated up to graduation. 12% respondents are educated up to post-graduation. Another 12.4% respondents having technical qualification or professional degree to their credit. Most of the respondents belong to supervisor category as 61.3% respondents falls into this group. 27.6% respondents are from management class and remaining 11.1% are from executive category respondents. The survey reveals that majority of respondents are having experience of 6–15 years as approximately 75% respondents indicated that they are associated with hospitality industry from last 6–15 years. Sample includes Luxury Hotels (26.7%) Mid-Scale Hotels (28.6%) Budget Hotels (42.9%) and Heritage Hotels (1.8%). The study signifies that sample is a combination of all the segment of employee having good experience with present organization.

Data summarized in Table 2 for various factors of employee motivation reveals that Work Autonomy factor secured the highest mean (m = 3.5565). It is followed by Job Recognition (m = 3.5196), Individual Organization Fit (m = 3.4624) Nature of Job (m = 3.4547), Recognition for the Work (m = 3.4401), Career Opportunity (m =

Table 2. Factors of employee motivation: a descriptive statistics

Factors of employee motivation	Total respondents (N)	Reliability(α)	Mean	SD
Nature of job		**0.803**	3.4547	.82597
I have a challenging and exciting job in my organization	217		3.6636	1.03289
My nature of job in the organization permits me to grow and develop new expertise	217		3.6221	.96469
Management Assign me the purposeful task	217		3.5392	.86044
Job recognition		**0.616**	3.5196	.66268
My organization cares about employee at personal as well as professional level	217		3.6866	.86252
I feel that I am contributing to my organization's overall objectives	217		3.5530	1.05769
My creativity and innovation are given due importance	217		3.3134	1.10280
Work appreciation motivates me to take more initiative	217		3.5253	.99096
Rewards and incentives		**0.812**	3.3585	.77096
My organization offers realistic long-term incentives	217		3.3041	1.07136
I am happy with the benefits offered by my organization	217		3.2811	1.07981
I get motivated by financial incentives to do my best work	217		3.4793	1.05444
Equal salaries for the workers are important motivator	217		3.5576	.86477
I am getting reward based on feedback from the management	217		3.1705	1.01537
Career opportunity		**0.677**	3.4393	.75871
Determining promotions by merit boosts my motivation	217		3.6452	.90184
A superior expresses a genuine interest in my career ambitions	217		3.2995	1.00354
My success is affected by the prospect of getting a promotion	217		3.3733	1.01097
Recognition for the work		**0.712**	3.4401	.71717
Providing recognition, credit, and acclamation motivates me to do my best	217		3.4516	.84366
In organization, I get recognition	217		3.3364	1.05067
The recognition I get from my manager directly motivates me to do my best	217		3.4194	1.01114

(continued)

Table 2. (*continued*)

Factors of employee motivation	Total respondents (N)	Reliability(α)	Mean	SD
In my organization, I know my work is noticed and valued	217		3.5530	.99917
Work Autonomy		**0.649**	3.5565	.73793
Flexibility at work positively influences my motivation	217		3.6221	1.00234
I have the flexibility to schedule my job	217		3.6406	1.03629
Flexible work climate encourages me to do my best every day at work	217		3.4793	1.14293
Employee participation in the decision-making is motivation	217		3.4839	1.04126
Individual organization fit		**0.715**	3.4624	.88410
My personality matches with organizational needs	217		3.5115	1.09337
My goal is aligned with organization's vision	217		3.4332	1.12892
I am given responsibility as per my ability and competence	217		3.4424	1.10036

3.4393) and Rewards and Incentives (m = 3.3585). Higher standard deviation of factor like Individual Organization Fit (SD = 0.8841) indicates that respondents view on this factor is heterogeneous. Looking at the measurement variable of employee motivation, "My organization cares about employee at personal as well as professional level", has scored highest mean of 3.6866 and standard deviation of 0.86252. This was followed by attributes like "I have a challenging and exciting job in my organization" with mean = 3.6636 and SD = 1.03289. attribute like I am getting reward based on feedback from the management with mean 3.1705 and SD = 1.01537 was ranked lowest by the respondents in the sample. Reliability of all the construct of employee motivation was found to be in the range of 0.616 to 0.812 that indicates the construct chosen for the employee motivation are consistent and reliable enough to go for further statistics analysis.

Table 3. Leadership approach: a descriptive statistics

	N	Reliability (α)	Mean	Std. deviation
Leadership approach (combined weightage)	217	.711	3.5452	.45529
Directing approach		0.716	3.7581	.58797
Management provides the employee with detailed instructions for routine work	217	5.00	3.8571	.78931
Management give employees specific goals and objectives	217	5.00	3.8065	.74496

(*continued*)

Table 3. (*continued*)

	N	Reliability (α)	Mean	Std. deviation
Management checks with workers regularly to keep track of personnel activities	217	5.00	3.8065	.74496
The management set an example by demonstrating the steps involved in doing the job	217	5.00	3.5622	.71818
Coaching approach		.719	3.5493	.56553
The leaders portray the role of management to their workers in a persuasive way	217	5.00	3.4516	.75689
Managers are trying to get people ready to take decisions without opposition	217	5.00	3.4101	.76527
Management positions workers on duty according to their ability to do the job	217	5.00	3.5438	.74484
Management admires employees for their good work	217	5.00	3.6359	.88255
Management gives advice a lot and frequently get input about how they are doing	217	5.00	3.7051	.74263
Facilitating approach		.745	3.6986	.58788
Management involves workers in the decision making that will influence their jobs	217	5.00	3.6728	.75083
Management tells its staff that they feel free to ask questions and discuss important issues	217	5.00	3.7558	.66686
Management holds regular meetings with workers to implement the latest concepts	217	5.00	3.6498	.70508
Management helps workers identify and promote the growth activities of their own growth activities	217	5.00	3.7189	.73236
Management listens without criticism or judgment about the staff issues and concerns	217	5.00	3.6959	.78152
Delegating approach		.738	3.7296	.69287
Management assigns wide-ranging duties to workers and expect them to handle these in full spirit	217	5.00	3.7143	.70804
Management expect staff to find and correct their own errors	217	5.00	3.7558	.76393
Management has the confidence in its employees and trust them	217	5.00	3.7189	.78130

Data presented in the Table 3 represent the descriptive statistics (mean and Standard Deviation) of various leadership approach like directing approach, coaching approach, facilitating approach and delegating approach generally used by the leaders in the organization. Descriptive statistics as calculated by SPSS software indicates that Directing Approach has scored highest mean of 3.7581 and SD = 0.58797. this was followed by

Delegating Approach with mean = 3.7296 and SD = .69287, Facilitating Approach with mean = 3.6986 and SD = .58788, Coaching Approach with mean = 3.5493 and SD = .56553. Higher standard deviation of factor like Delegating Approach (SD = .69287) indicates that respondents view on this factor is heterogeneous. Looking at the measurement variable of leadership approach, "Management provides the employee with detailed instructions for routine work" with m = 3.8571 and SD = 0.78931 has scored highest followed by "Management give employees specific goals and objectives" with mean = 3.8065 and SD = 0.74496. Attributes like Managers are trying to get people ready to take decisions without opposition with mean = 3.4101 and SD = 0.76527was ranked lowest by the respondents in the sample. Reliability of leadership approaches were in the range of 0.716 to 0.745 that indicates the construct chosen for the leadership approach are consistent and reliable enough to go for further statistics analysis.

Table 4. Organizational performance: a descriptive statistics

Description	Total respondents (N)	Reliability (α)	Mean	SD
	217	.725	3.5452	.45529
Operating cost of organization has reduced significantly			3.7880	.77653
Average occupancy rate has increased			3.5760	.83054
Organization has increased the revenue from innovative alternative revenue sources			3.3871	.82631
Employee turnover has reduced thus leading to reduction in recruitment cost			3.4793	.87707
Average length of stay of customer has improved			3.5253	.81681
Customer satisfaction has increased			3.5609	.94070
Market penetration rate has improved			3.4435	.94098
Customer defection rate has reduced significantly			3.5739	.84172
Better market goodwill			3.6217	.69336
Improvement in the brand equity			3.5304	.85979
Valid N (listwise)				

Data presented in the Table 4 represent the descriptive statistics (mean and Standard Deviation) of various measurements related to performance of hospitality industry as rated by respondents. It is observed that measurement variable like "Operating cost of organization has reduced significantly" was ranked highest with mean = 3.788 and SD = 0.77653. This was followed by attributes like Better market goodwill with mean = 3.6217 and SD = 0.69336 and Average occupancy rate has increased with mean = 3.576 and SD = 0.83054. The attributes like organization have increased the revenue from innovative alternative revenue sources with mean = 3.3871 and SD = 0.82631 has scored lowest among all variable.

Table 5. Relationship between factors of employee motivation and organizational performance: direct effect

Model		Unstandardized Coefficients		Standardized Coefficients		
		B	Std. Error	Beta	t	Sig.
1	(Constant)	1.468	.259		5.666	.000
	Employee Motivation	.602	.074	.483	8.089	.000
R= .483		R^2 = .233				
F = 65.437		P =.000		Std. Error of the Estimate = .39656		

a. Dependent Variable: Organizational Performances

Regression analysis was carried out to measure coefficients of linear relationship between employee motivation and organizational performance. Combined factors mean (Nature of Job, Job Recognition, Rewards and Incentives, Career Opportunity, Recognition for the Work, Work Autonomy, and Individual Organization Fit) were used in analyzing the employee motivation. Similarly, combined mean of Directing Approach, Coaching Approach, Facilitating Approach and Delegating Approach were calculated for assessing Leadership approach. Combined mean of 10 measurement variable of perceived organizational performance were calculated using SPSS software and then further regression analysis was carried out. Table 5 shows the results of regression analysis. Impact of Employee Motivation on Organization performance was found significant ($f = 65.437$, $p = .000^b$, $t = 8.089$, $p = .000$) and contributed 23.3 ($R^2 = 0.233$) to the organizational performance. The results revealed that the beta values for employee motivation are 0 .602 and it has significant effect on organization performance (Table 5).

Indirect of Effect of Employee Motivation on Organizational Performance: Regression Analysis

For assessing the indirect effect of employee motivation on organizational performance, regression analysis was carried out in two phases. In the first stage, regression analysis was carried out to assess linear relationship between combined effect of various motivational practices (Nature of Job, Job Recognition, Rewards and Incentives, Career Opportunity, Recognition for the Work, Work Autonomy, and Individual Organization Fit) was taken as independent variable on leadership approach (Directing approach,

Table 6(a). Impact of employee motivation of leadership approach

Model		Unstandardized Coefficients B	Unstandardized Coefficients Std. Error	Standardized Coefficients Beta	t	Sig.
1	(Constant)	1.070	.207		5.168	.000
	Employee Motivation	.757	.059	.655	12.715	.000
R = .655		R^2 = .429				
F = 161.670		P = .000		Std. Error of the Estimate = .31699		

a. Dependent Variable: Leadership Approach

Table 6(b). Impact of leadership approach on organizational performance

Model		Unstandardized Coefficients B	Unstandardized Coefficients Std. Error	Standardized Coefficients Beta	t	Sig.
1	(Constant)	1.087	.215		5.064	.000
	Leadership Approach	.668	.058	.619	11.559	.000
R = .619		R^2 = .383				
F = 133.607		P = .000		Std. Error of the Estimate = .35568		

a. Dependent Variable: Organizational Performance

Coaching Approach, Facilitating Approach, Delegating Approach) was taken as dependent. In the second step, Leadership approach was taken as independent variable and organizational performance was taken as dependent variable. Result of regression outcome as presented in the above Table 6(a) and 6(b) indicates that employee motivation has significant impact (Beta (β) = .757, p = .000) on leadership approach of employee. R^2 Value (0.429) shows independent variables are moderately contributing towards change in dependent variable. In the second step regression analysis was carried out between leadership approach and organizational outcome. Outcome indicates that leadership have significant impact (Beta (β) = .668, p = .000) on organizational performance. R^2 Value (0.383) depicts independent variables are significantly contributing towards change in dependent variable (organizational performance).

Further, Sobel test was carried out to analyze influence of mediating variable. Outcome is presented in the table below (Table 8):

Table 7(a). Influence of mediating variable using Sobel test (A)

	Input		Test statistics	Standard error	P-value
a	.757	Sobel test	8.57072724	0.05900036	.000
b	.668	Aroian test	8.55634772	0.05909951	.000
Sa	.059	Goodman test	8.5851795	0.05890104	.000
Sb	.058				

Table 7(b). Influence of mediating variable using Sobel test (B)

	Input			
t_a	12.715	Sobel test	8.55299775	.000
t_b	11.559	Aroian test	8.53855165	.000
		Goodman test	8.56751741	.000

Table 8. Testing of hypothesis

Sl. no	Hypothesis	Beta (β)	t-test	R^2	Remarks
1	Employee motivation have positive relationship with organizational performance	.602	8.089	.233	Significant
2	Employee motivation have positive influence on leadership	.757	12.715	429	Significant
3	Leadership practices in the organization have positive influence on organizational performance	.668	11.559	.383	Significant
4	Leadership practices in the organization mediates the relationship between employee motivation and organizational performance	Refer Table 7(a) and 7(b)			Significant

Sobel test, Aroian test and Goodman test were conducted to evaluate the effect of the mediating variable (leadership approach) of an independent variable (factor of employee motivation) on a specific dependent variable (organizational performance). In general, the case of mediation can be said to be if (1) the Independent Variable has a significant effect on the mediator; (2) the Independent Variable has a substantial effect on the Dependent Variable in the absence of the mediator; (3) the mediator has a large special influence on the Dependent Variable. Those criteria can be used to assess whether mediation takes place informally or not, while the statistically based methods for standardized study mediation have been used by Mac Kinnon and Dwyer (1993) and MacKinnon, Warsi and Dwyer (1995). Sobel test revealed that p-values are less than.05, indicating that the two-size z-test is greater than 1.96, and therefore the null hypothesis is dismissed and supports the alternative hypothesis that leadership approach mediates the relationship between employee motivation and the organizational performance of hospitality sector industries in Uttarakhand (India).

4 Discussions and Conclusion

The present research explored various factors of employee motivation and their effect on organizational performance and explore whether leadership approach mediates the relationship between employee motivation and organizational performance. From the results analyzed it was found that employees rated Work Autonomy (m = 3.5565) as an important factor of their motivation. The present finding supports the study of Gellatly and Gregory (2001) who confirm that majority of the employee perceives that increase in the job performance will be due to job autonomy. This research also supports the work of Morrison et al. (2005) revealed job autonomy to be significant factor in employee's job satisfaction. High autonomy influences the view that salespeople have of their work. We will fail, because their efforts, acts and decisions will decide the outcomes of their work. This condition affects their judgment behavior in performing the job which leads towards improved job satisfaction as well as work performance and contributes to better performance of the organization. The regression study shows that employee motivation positively affects the organization's performance and leadership strategy mediates the relationship between employee motivation and efficiency in organization. This study also supports the previous findings of (Aksu 2012; p. 5, Koçak and Özüdoğru, p. 85; Türker 2014, p. 800; Gopal and Chowdhury 2014, p. 7) it is certain that leadership style as well as employee motivation are considered as an organizational tool for improving organizational performance specially in hospitality organization that is primarily dominated by human centric industry. An improved leadership style is perceived to contribute positively to the growth of organizations. Leadership styles in the sampled organization (hospitality) is mostly ranked by carryout descriptive statistics calculated by SPSS software. Result indicates that Directing Approach has scored highest mean of 3.7581 and SD = 0.58797. this was followed by Delegating Approach with mean = 3.7296 and SD = .69287, Facilitating Approach with mean = 3.6986 and SD = .58788, Coaching Approach with mean = 3.5493 and SD = .56553. This means that the performance level which is encouraged by ability to direct employee and facilitating them with proper support system. Result also confirms that organizational performance is influenced by leadership approach as well as their behavior.

Implications for Theory and Practice

Most of the hospitality sector organizations place greater emphasis on raising employee motivation through permutation and combination of different human resource management practices with the intention generate meaningful results for both employees as well as organization. The effect of employee motivation and leadership practice on organizational performance is significant as most of the researcher pointed out. The work autonomy fit between personal and organization, suitable financial and non-financial incentive, job recognition and positive psychological resources has gained more significance. This study suggests that improved leadership approach with suitable recruitment policies ensure individual organization fit and giving more work freedom are some of the factors to improve employee motivation and contribute to enhanced organizational performance. According to our findings, designing suitable human resource management strategy suiting the employee work autonomy and aligning employee expectation with organization goal is vital for organizational performance. In applying these human resource strategies, leadership approach may play an active role for recruiting staff whose personality are suiting to the organization standard so that employee can be motivated with minimum resistance and take greater participation towards organizational performance.

Theoretical Contribution

This study offers theoretical contribution to develop organization strategies for managerial solutions of employee motivation through exploring dimensions of employee motivation through leadership approach for better performance. The hospitality organizations aiming to enhance their performance must focus on various components of employee motivation. Result supports that organization should invoke leadership orientation with the intention of enhancing employee motivation through problem solving and cognitive as well as analytical experience (Fiss 2007). Because qualitative analysis involves comprehensive consideration of context and social activities, researchers would tend to think in terms of combinations and configurations. In terms of "causal ingredients" researchers will also consider the causal conditions, that interact to obtain the desired outcome. A summary of how things are going to happen can also be given by this interest in combinations of factors. Configurational approach thus recommends that organizations are best understood as clusters of interconnected structures and activities than as modular or loosely linked entities whose components can be understood in isolation (Fiss 2007, 1180).

Practical Implication

The study provides some practical solutions based on an evidential understanding of a leadership approach that helps empower workers and leads to improved organizational efficiency. It is necessary for organization's management to understand the actions of leaders based on characteristics and abilities (Lussier and Achua 2010). There are two distinct leadership styles based on the individual's attitude. Leaders with elevated structure and low thinking. To make choices, these leaders use unilateral contact. On the other side, with high thinking and low structure, there are leaders. The communication here is bilateral and both the leader and the team make the decisions. This research would also help us become cognizant of all the theories of motivation. There is nothing constant and the world of business is no exception. A leader should be able to efficiently handle the transition. There could be too much strain, delays in completing the job, barriers, etc. Such unforeseen incidents should be taken care of by the organization and the members. This study allows us to better understand the attitudes of workers towards motivational practices and the expected motivational approach to leadership to better performances and future success.

Limitations and Future Research

There are some drawbacks to the present research that must be discussed. The study's first drawback would appear to derive results from a cross-sectional nature of sample. And final conclusions cannot be made regarding the causal relationship. These tests should be complemented by a variety of methods which could make the findings more important. Longitudinal study designs are required to examine the proposed model for achieving a more acceptable causality. The second downside is that all results are focused on qualitative and self-reporting and therefore the findings of the traditional variance approach can be distorted. Although statistical analysis showed that there was no significant concern about the variance of a specific system, potential effects of variance in the standard method may influence the findings. Future studies should collect data from several sources, to prevent such a problem. Relevance can be increased by drawing the result from a larger population. In addition, in future research, it is important to explore further factors of employee engagement and leadership style for building psychological resources for employees that are helpful in enhancing efficiency and organizational performance.

References

Aksu, G.: The effect of team leaders on employees' motivation: call center review. Acad. Look Mag. (32), 1–21 (2012)

Asrar-ul-Haq, M., Kuchinke, K.P.: Impact of leadership styles on employees attitude towards their leader and performance: empirical evidence from Pakistani banks', Future Bus. J. **2**(1), 54–64 (2016)

Baldoni, J.: Motivation Secrets. Great Motivation Secrets of Great Leaders (2005). http://govleaders.org/motivation_secrets.htm

Bartol, K.M., Martin, D.C.: Management. McGraw Hill, New York (1998)

Bouckenooghe, D, Zafar, A, Raja, U.: How ethical leadership shapes employees' job performance: the mediating roles of goal congruence and psychological capital. J. Bus. Ethics **129**(2), 251–264 (2015)

Çetin, I.: Motivation and its impact on labour productivity at Hotel business a conceptual study. Int. J. New Trends Arts, Sports Sci. Educ. **2**(1), 70–79 (2013)

Cross-cultural analysis of the motivational antecedents of hotel workers in Mauritius andAustralia. J. Manage. Dev. **24**(3), 253–266 (2005)

Fishbach, A., Choi, J.: When thinking about goals undermines goal pursuit. Organ. Behav. Hum. Decis. Process. **118**, 99–107 (2012)

Fiss, P.C.: A set-theoretic approach to organizational configurations. Acad. Manage. Rev. **32**(4), 1180–1198 (2007)

García-Morales, V.J., Jiménez-Barrionuevo, M.M., Gutiérrez-Gutiérrez, L.: Transformational leadership influence on organizational performance through organizational learning and innovation. J. Bus. Res. **65**(7), 1040–1050 (2012)

Gellatly, I.R., Irving, P.G.: Personality, autonomy and contextual performance for managers. Hum. Perform. **43**(3), 231–245 (2001)

Gobal, R., Chowdhury, G.: Leadership styles and employee motivation: an empirical investigation in a leading oil company in India. Int. J. Res. Bus. Manage. **2**(5), 1–10 (2014)

Hancout, D.C.: Relationship between Transformational leadership and Organisational Performance in a Largest Public Sector Company in Canada, Proquest Information and Learning Company, Annalbor MI (2005)

Judge, T.A., Ilies, R.: Relationship of personality to performance motivation: a meta-analytic review. J. Appl. Psychol. **87**(4), 797–807 (2002)

Judge, T.A., Piccolo, R.F.: Transformational & transactional leadership: a meta-analytic test of their relative validity. J. Appl. Psychol. **89**(5), 755–768 (2004)

Keller, R.T.: Transformational leadership, initiating structure & substitutes for leadership: a longitudinal study of research & development project team performance. J. Appl. Psychol. **91**(1), 202–210 (2006)

Kieu, H.Q.: Leadership Styles and Organizational Performance: A Predictive Analysis: ERIC (2010)

Kingir, S., Mesci, M.: Factors that affect hotel employees' motivation: the case of Bodrum. Serbian J. Manage. **5**(1), 59–76 (2010)

Koçak, R.D., Özüdoğru, H.: Effects of managers' leadership features on employee motivation and performance: an application in public and private hospitals 2012. J. Faculty Commer. Tourism Educ. 1 76–88 (2012)

Koçak, R.D., Özüdoğru, H.: Effects of managers' leadership characteristics on employee motivation and performance: an application in public and private hospitals. J. Commer. Tourism Educ. Faculty (1), 76–88 (2012)

Lussier, R.N., Achua, C.F.: Leadership: Theory, Application, & Skill Development, 5th edn. South-Western, Mason, OH (2010)

McGrath, G.R., MacMillan, I.C.: Entrepreneurial Mindset: Strategies for Continuously Creating Opportunity in an Age of Uncertainty. Harvard Business School Press Books, Brighton (2000)

Meyer, G.D., Heppard, K.A.: Entrepreneurial Strategies: The Dominant Logic of Entrepreneurship. Irwin University Books, New York (2000)

Morrison, D., Cordery, J., Antonia, G., Roy, P.: Job design, opportunities for skill utilization and instrinsic job satisfaction. Eur. J. Work Organ. Psychol. **14**(1), 59–79 (2005)

Motivation needs in Kenya's hotel sector. Afr. J. Hospitality, Tourism and Leisure **4**(1), 1–10

Maithil, N., Chaubey, D.S., Gupta, D.: Impact of organization culture on employee motivation and job performance, IJRCM. ISSN 0976–2183, vol. 3, Issue 5, May 2012

Purcell, J., Kinnie, N., Hutchinson, S., Rayton, B., Swart, J.: Understanding the people & performance link: unlocking the black box. Research Report, Chartered Institute of Personnel and Development (2004)

Riedle, D.: Transformational vs. transactional leaders: how different leadership behaviors and communication styles affect levels of employee motivation in the financial industry (Order No. 1591122). http://search.proquest.com/docview/1696043546?accountid=142908 (2015)

Ross, D.L.: Perceived job characteristics and internal work motivation: Exploratory (2005)

Danişman, Ş.: The Effect of Leadership on Organizational Performance, Leadership and Organizational Outcomes, pp. 143–168 (2015)

Singh, S.K., Tiwari, V.: Relationship between motivation and job satisfaction of the white-collar employees: a case study, SMS Varanasi, 11 December, vol. VII(2), pp. 31–39 (2011)

Touré-Tillery, M., Fishbach, A.: The end justifies the means, but only in the middle'. J. Exp. Psychol. Gen. **141**, 570–583 (2012)

Touré-Tillery, M., Fishbach, A.: How to measure motivation: a guide for the experimental social psychologist. Soc. Pers. Psychol. Compass **8**(7), 328–341 (2014)

Türker, M.: (Employee Perceived Leadership Styles and Motivation Levels Relationship, 2nd Organizational Behavior Congress, Papers (2014)

Wasike, C.K., Ndivo, R.M.: Efficacy of motivation strategies in addressing employee (2015)

Yukl, G.: Leadership in Organizations, 5th edn. Prentice-Hall, Englewood Cliffs (2002)

Fair Exposure: A Multi-stakeholder Personalized Recommendation System Based on Multi-objective Optimization

Rahul Shrivastava(✉), Dilip Singh Sisodia, and Naresh Kumar Nagwani

Department of Computer Science and Engineering, National Institute of Technology Raipur, G E Road, Raipur 492010, Chhattisgarh, India
{rshrivastava.phd2018.cs,dssisodia.cs,nknagwani.cs}@nitrr.ac.in

Abstract. The conventional way of recommendation is exclusively focused on end-user personalization and accuracy improvement. Nowadays, business models are adapting a multi-sided fairness policy, and the multi-sided fairness could be achieved with a multi-stakeholder recommendation model. Recent studies suggested that the existing end-user focused recommendation model lacks in justifying other stakeholders' needs and does not create any personalized recommendation environment. Thus, this research proposes a multi-objective optimization-based personalized recommendation framework suitable for preferences for every stakeholder in the system. Particularly this work considers the two stakeholders in the proposed recommendation model producer and system. We design two dedicated objective functions to address the preferences of producer and system stakeholders. The producer's objective function is to enhance the producer's exposure in the recommendation list, and the objective function of the system is intended to observe balance and fair recommendation. Next, to optimize the conflicting utilities of stakeholders, we propose a stakeholder optimization-based recommendation framework. The proposed model aims to improve utility gain and establish a balanced trade-off recommendation solution between stakeholders. The proposed model is evaluated based on cumulative utility gain against the baseline algorithms. The evaluation result describes the significant improvement in cumulative utility gain when the standard algorithms are altered with proposed objective functions.

Keywords: Multi-stakeholder recommendation · Multi-objective recommendation · Multi-sided fairness · Personalization

1 Introduction

The personalized recommendation system can manage large amounts of information flow in many application domains, such as e-commerce [1]. For any recommendation to be valid end-user personalization is an important criterion [2]. In the recent business application domain, the customers are not the only valid stakeholders [3]. Consider a real-world example of a movie or education recommendation where other stakeholders

L. Garg et al. (Eds.): ISMS 2020, LNNS 303, pp. 202–213, 2022.
https://doi.org/10.1007/978-3-030-86223-7_18

are equally important such as in an educational setting beyond student teachers, and college authorities are the primary stakeholders [4, 5]. Side information modeling [6, 7] or content-based similarity [8] approaches exclusively focus on end-user accuracy improvement. The reciprocal recommendation [9, 10], such as the job recommendation [11] and dating application, is the similar real-world application of multi-stakeholder. The multi-sided fairness concept seeds the ground for multi-stakeholder recommendation [12].

The term multi-sided fairness indicates the footprint of multi-stakeholder in the past. In recent studies, the utility function-oriented approach has shown promising capabilities in justifying different stakeholder's needs in the system. The work proposed in [13] designs the dedicated utility functions for the student, instructor, and system as a valid stakeholder in the educational setting. The utility optimization-based model proposed in [14] Generates the trade-off between conflicting utilities of stakeholders. The traditional memory-based or model-based collaborative filtering techniques [15, 16] are mainly focusing on improving recommendation accuracy. The utility functions targeting only end-user preferences may achieve better accuracy but lacks in novel and diverse item recommendation [17, 18].

This research produces a novel contribution to a multi-stakeholder recommendation by proposing a stakeholder optimization-based recommendation model. Recent studies lack consideration of the producer's growing needs and the system's objective of achieving balance and fairness. Hence this work proposes dedicated objective functions for two stakeholders of the recommendation process. Next, the conflicting preferences of the stakeholders are optimized using multi-objective optimization techniques. The major contributions are summarized as follows:

- We propose a stakeholder objective optimization-based recommendation model with two aims: first, to establish a balanced and fair trade-off recommendation solution and improve cumulative utility gain.
- We propose two objective functions for addressing the preferences of stakeholders. The objective functions can enhance producer stakeholder's exposure in the recommendation list and achieve suitable exposure count, unique user count, and recommendation count respectively for producer and system.
- The proposed model is evaluated based on the cumulative utility gain with other baseline algorithms. The standard algorithms are modified with proposed objective functions to compare the utility values.

The rest of the section organized as follows in Sect. 2 presents the literature review; Sect. 3 describes the proposed stakeholder optimization-based recommendation model. Further, Sect. 4 presents experimental results, and Sect. 5 provides utility evaluation result discussion, and finally, Sect. 6 concludes the paper.

2 Related Work

The multi-objective recommendation model used to provide a non-dominated Pareto optimal solution to the end-users conflicting objectives; the same could be extended towards multi-stakeholder recommendations [19, 20]. The multi-objective recommendation model targeted towards considering beyond accuracy objectives like novelty [18], serendipity, diversity [21]. These models are extended towards beyond end-user consideration in the multi-stakeholder recommendation. A diverse and novel item recommendation based approach described in [19] develops a multi-objective recommendation model. The multi-objective recommendation enhances the economic stability of e-commerce. A multi-route travel planning approach is proposed to improve the economic profit with tourism [22]. A probability-based multi-objective recommendation model proposed in [20] considers diversity as a non-accuracy indicator objective function.

The term multi-sided fairness in business initiated the multi-stakeholder recommendation culture [12]. Further, the multi-stakeholder recommendation is implemented as a group of users in a group recommendation system [5, 23]. But, strict multi-stakeholder recommendation considers the group of users as a single stakeholder. After the group recommendation, two stakeholder-based reciprocal recommendations are the next crucial development in multi-stakeholder recommendation [9, 11]. Formally, the multi-stakeholder recommendation patterns introduced in [20] formally describe three stakeholders: consumer, producer, and system, and present active and passive consumer and system personalization. Multi-stakeholder recommendation ideas are explored in various application domains, like music [24] educational learning [25]. Various applications and challenges in the area of multi-stakeholder recommendations are investigated in [26]. Stakeholder's objectives are necessary to justify the recommendation system [27]; the system also has its concern like fairness, a long-tail recommendation incorporated in the recommendation system [4]. Still, in recent literature, we found many application areas where multi-stakeholders impact should be explored. Even in existing movie recommendations, there is a need to define objective functions to address stakeholder objectives, and the recommendation system's beyond accuracy objectives.

The utility optimization-based approach proposed in [14] optimizes stakeholder preferences while taking the help of a multi-criteria rating [28] to build an expectation vector as a stakeholder's utility. End-user focused utility on improving the economic consideration of user proposed in [29]. Utilities of providers are presented as a constraint in multi-stakeholder recommendations [30].

The majority of multi-stakeholder recommendations with utility functions approach satisfying stakeholder objectives, but a very limited approach is found to obtain utilities from top-n recommendation. Incorporating beyond accuracy objectives is another significant update that can boost the utility function's capabilities. Hence this research proposes dedicated smart, objective functions for each stakeholder producer and the system. The proposed objective function improves the various utility count, such as exposure count, unique user count, and item's recommendation count.

3 Proposed Multi-stakeholder Recommendation Model

This section describes the different modules and steps involved in the proposed multi-stakeholder recommendation model. In the proposed model, we consider the conflicting relationship between two stakeholders in the movie recommendation system: Producer and System. Both considered stakeholders have their inconsistent preferences, such as a specific producer concentrates on a movie's recommendation from his production house irrespective of the system's objective. In contrast, the system as a stakeholder can achieve balance in stakeholder preferences and maximize profit gain. The proposed multi-stakeholder recommendation framework establishes the trade-off between stakeholders. This section first describes the dataset and further presents the proposed objective functions associated with star cast and producer. Next, A multi-objective optimization using an evolutionary algorithm is presented for the conflicting preferences of stakeholders.

3.1 Proposed Objective Functions

This section presents the algorithms explaining the objective functions for producer and system. The producer's objective function is exposure-oriented and focusing on improving the exposure of the producer's item. The system's objective function is to justify every stakeholder's need in the recommendation framework and improve aggregate profit.

3.2 Exposure-oriented Algorithm for Producer

In this research, we propose an exposure oriented algorithm for broadcasting the preferences of producer stakeholder. The proposed exposure oriented algorithm modifies the recommendation list to incorporate beyond user and accuracy capabilities in the recommendation list. Algorithm 1 describes the exposure oriented algorithm where we first generate a top-n recommendation list and modifies the recommendation list based on exposure count. The product's appearance counted for specific producer stakeholders and based on appearance count, the product's exposure rate is decided. The products with low exposure rate or low appearance count are promoted in the recommendation list. First, the predicted user-item rating metrics generated (refer step-1 algorithm-1) than for each user's respective recommendation list is generated (refer step 2–3). The exposure count value is calculated for each item in a recommendation list of every user (refer step 4–7) and producer information being fetched. The recommendation list is updated according to the exposure count value to promote the producer's low exposure count items.

Algorithm 1: Exposure-oriented objective function for enhancing producer exposure

Input: *User item original rating metrics* $\boldsymbol{r_{u,i}}$

Output: *Personalized Producer Exposure utility*

1 Compute user-item interaction rating metrics with predicted ratings $\hat{\boldsymbol{r}}_{u,i}$ using User-based or Item-based collaborative filtering technique.

2 ***for*** *each user* $\boldsymbol{U_i}$:

3 $\quad$ *Generate top* $-$ *n recommendation* ***list*** $\boldsymbol{L} = \{\boldsymbol{L_1}, \boldsymbol{L_2}, \ldots \boldsymbol{L_u}\}$

4 ***for*** *each user* $\boldsymbol{U_i}$:

5 $\quad$ ***for*** *each recommendation list* $\boldsymbol{L_u}$ *of* $\boldsymbol{U_i}$:

6 $\quad\quad$ ***for*** *each iten* $\boldsymbol{k}$ *in* $\boldsymbol{L_u}$:

7 $\quad\quad\quad$ ***Update*** *exposure count* $\boldsymbol{Exp_{count}(k)} += \boldsymbol{1}$

$\quad\quad$ ***fetch*** *the producer* P *where* $k \in P$

8 ***for*** *each producer* P :

9 $\quad$ *fetch movie item* M_i *produced by* P *such that* $M_i \in P$

10 $\quad$ *if* $\boldsymbol{Exp_{count}}(M_i) \leq \boldsymbol{0}$ ***and*** $M_i \notin L$

11 $\quad\quad$ ***Include*** M_i *in the recommendation list* $\boldsymbol{L}$

12 *Update* $\boldsymbol{Exp_{count}}(K)$ *with each timestamp*

13 *Normalized Exposure count/fitness value(* f_1

$$f_1 = NExp_{count} = \frac{\mathbf{1}}{|U||L|} \sum_{L_u \in L} \sum_{k \in L_u} Exp_{count}(k)$$

3.3 Balance and Fair Algorithm for System

In this research, we propose a balanced and fair objective function for system stakeholders. The objective function presented in algorithm two is focused on achieving three major objectives: (a) maintain fair and balanced stakeholder exposure count (Refer algorithm 2, Step 4–11), (b) create personalization for stakeholder by updating unique user count value (Refer algorithm 2, Step 13–15) and (c) fair towards the recommendation of every item using item recommendation count (Refer algorithm 2, Step 17–19).

Algorithm 2: Balanced and Fair objective function for System stakeholder

Input: *User item original rating metrics* $\boldsymbol{r_{u,i}}$

Output: *Aggregate system utility value*

Compute user-item interaction rating metrics with predicted ratings $\hat{r}_{u,i}$ using User-based or Item-based collaborative filtering technique.
for each user $\boldsymbol{U_i}$:
    Generate top − n recommendation **list** $\boldsymbol{L} = \{L_1, L_2, \ldots L_u\}$
#For Stakeholders
for each stakeholder $\boldsymbol{S_i}$:
    for each recommendation list $\boldsymbol{L_u}$ *of* $\boldsymbol{U_i}$:
        for each item $\boldsymbol{k}$ *in* $\boldsymbol{L_u}$:
            if $\boldsymbol{k} \in \boldsymbol{S_i}$:
                Update stakeholder exposure $\boldsymbol{Stack\_Exp_{count}(S_i) += 1}$
for each stakeholder $\boldsymbol{S_i}$:
    If $\boldsymbol{Stack\_Exp_{count}(S_i) < 0}$
        Promote *stakeholder's item into the recommendation list.*
#Providing Unique user count to the stakeholders to create personalization
for each recommendation list $\boldsymbol{L_u}$ *of* $\boldsymbol{U_i}$:
    for each iten $\boldsymbol{k}$ *in* $\boldsymbol{L_u}$:
        Update unique user count(UUC) for which $k \in \boldsymbol{L_u}$: $\boldsymbol{UUC_{Count}(k)+}$
Create a personalized list of target users for
for each iten $\boldsymbol{k}$ *in* $\boldsymbol{L_u}$
    Compute Item Recommendation count $\boldsymbol{IRC(k) += 1}$
  The cumulative aggregate system fitness value

$$f_2 = NSystem_{count} = \frac{1}{|U|.|L|}\sum_{L_u \in L}\sum_{k \in L_u}(StackExp_{count}(S_i) + UUC_{Count}(k) \cdot IRC(k))$$

3.4 Proposed Stakeholder Objective Optimization-Based Recommendation Model

This research proposed a multi-objective optimization-based recommendation model to optimize producer and system stakeholders' conflicting objectives. The proposed model is capable of establishing a trade-off between both stakeholders. Algorithm 3 here explains the evolutionary algorithm-based optimization steps. Here we evaluate the fitness values of both fitness function (f_1,f_2) mentioned in algorithm 1 and 2. The algorithm continues to be executed and generating new recommendation solutions until the maximum number of generations achieved.

Algorithm 3: ***Proposed Stakeholder objectives Optimization-based Recommendation Framework***

Input: Top-n Recommendation list (RL) generated using Neighbourhood based collaborative filtering model.

Output: Multi-stakeholder objectives-based trade-off top-n recommendation Solution

Parameter Initialization: Stakeholders, genetic operators like population, selection, crossover a and mutation technique, number of objectives, number of a variable with lower and upper bound.

```
Chromosome Representation
(Natural encoding-based representation of Initial population)
for each stakeholder S do
    for each user U do
        Fitness evaluation of producer stakeholder objective function
        (using algorithm 1)
        Evaluating the fitness of system stakeholder objective function
        (using algorithm 2)
While the maximum number of generation reached:
        Continue genetic operation like selecting parents, crossover, and new
        offspring generation.
        Perform two objective optimizations using NSGA-II.
        Generate new Recommendation Solution
return trade-off recommendation solution
```

4 Experiments

This section describes the dataset, experimental results related to the Pareto front generated to establish a trade-off between conflicting objectives. To analyze the Pareto front result and evaluate the utility gain fifty independent execution of the proposed stakeholder optimization-based recommendation model performed over the movie lens 100k dataset. The experiments were conducted on a 64-bit machine with i7 10th generation @2.3 Ghz with 8 GB RAM in the windows environment.

4.1 Dataset

This research used a movie lens dataset consisting of 100,000 ratings ranging from 1–5 from 943 users and 1682 movies. Each user has rated at least 20 movies. The necessary contextual information age, gender, occupation, and zip contained by the dataset. The dataset was collected through the grouplens website, which made the rating dataset the URL referred to as (https://grouplens.org/datasets/movielens/). The stakeholder information related to the producer and star cast of the movie crawled from IMDB.

4.2 Trade-off Recommendation Solution: Pareto Front

Trade-off recommendation solution presented using a Pareto front in Figs. 1 and 2. Both figures describe a non-dominated Pareto optimal solution to the conflicting objective scenario of two different stakeholders. The generated Pareto front indicates the established trade-off so that both utilities can grow without hurting others. The non-dominated solutions result from the mating, i.e., selection, crossover, and survival of individuals using NSGA-II [31].

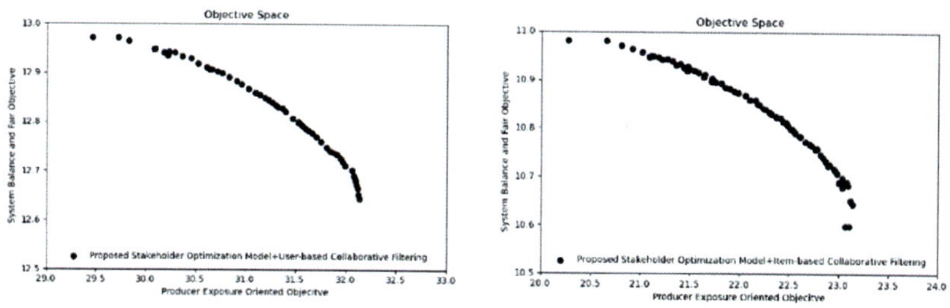

(a) (b)

Fig. 1. (a) Pareto front proposed stakeholder optimization + User based Collaborative filtering (b) Pareto front proposed stakeholder optimization + Item based Collaborative filtering

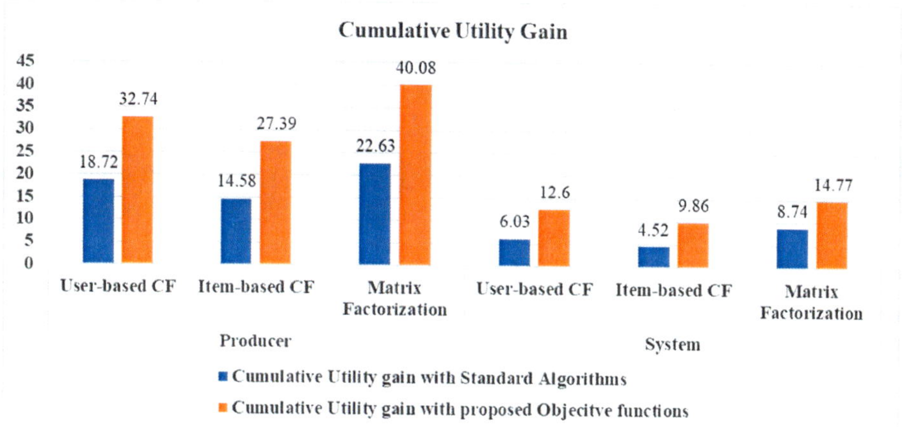

Fig. 2. Cumulative utility evaluation for different stakeholders against baseline algorithm

4.3 Evaluation

To validate the proposed model's performance, we altered the standard algorithms to incorporate the proposed objective functions. The cumulative utility improvement or loss of the proposed stakeholder objective optimization model is compared with baseline algorithms like user-based collaborative filtering [32], item-based collaborative filtering

[33], basic matrix factorization [34]. We observed that producer and system stakeholder's proposed objective functions significantly improve the cumulative utility gain with the evaluation results. The improvement is that algorithm 1 keeps track of producer stakeholder's appearance in the recommendation list and enhances the chances of getting recommended with exposure count value. Similarly, in algorithm two, the system stakeholder monitors the recommendation activity closely based on three objectives. First, the system stakeholder is improving stakeholders exposure count in a fair way, secondly it helps stakeholder with personalized target unique user count and third the system is balancing the recommendation count of every item. The Table 1 summarizes the cumulative utility gain or loss results. The above base line algorithms are altered for integrating the proposed objective functions for both the stakeholders, and utility values are comapred for both standard algorithm and modified algorithm with proposed objective functions.

5 Discussion

This research specifically considered beyond user stakeholders producer and system and designed a stakeholder optimization-based recommendation model. To track stakeholder's preferences and improve the utility gain for both stakeholders, we present two objective functions; the first one is for producer stakeholders and intended to enhance individual producer stakeholders' exposure in the recommendation list. The second objective function is directed towards balancing the exposure of every producer in the recommendation list. It also manages the unique user count and Item recommendation count to achieve personalization for stakeholders and provide equal opportunity to every item in the recommendation list. The objective functions are incorporated into the recommendation algorithm, which generates a top-n recommendation list.

Further, the top-n recommendation list is modified with evolutionary algorithm-based multi-objective optimization techniques to generate a new trade-off recommendation solution. The Pareto front presented in Fig. 1(a) and (b) is the Pareto optimal solutions concerning user-based and item-based collaborative filtering algorithm. The evaluation results given in Table 1 describe the cumulative utility gain with a standard baseline algorithm and a modified baseline algorithm with the objective function. The proposed stakeholder optimization-based recommendation model incorporated with proposed objective functions for stakeholders achieves a better optimal trade-off recommendation solution with significant improvement in the cumulative utility gain.

Table 1. Utility gain for different stakeholders against the baseline algorithm

Stakeholders	Algorithm	Cumulative utility gain with standard algorithms	Cumulative utility gain with proposed objective functions	Remark
Producer	**User-based CF**	18.72	32.74	**42.82% (Improvement)**

(continued)

Table 1. *(continued)*

Stakeholders	Algorithm	Cumulative utility gain with standard algorithms	Cumulative utility gain with proposed objective functions	Remark
	Item-based CF	14.58	27.39	**46.76% (Improvement)**
	Matrix factorization	22.63	40.08	**43.53% (Improvement)**
System	**User-based CF**	6.03	12.60	**52.14% (Improvement)**
	Item-based CF	4.52	9.86	**54.15% (Improvement)**
	Matrix factorization	8.74	14.77	**40.82% (Improvement)**

6 Conclusion

The recommendation system shifts focus towards other stakeholders of the recommendation model to meet the current business need. The end-user focused approach of objective optimization or utility enhancement has limited capability in justifying the stakeholder's requirement. With the same concern, this research presents a specific stakeholder optimization-based recommendation model and considers two movie producers and systems as valid stakeholders of the system. The first objective of this research is to design dedicated objective functions for addressing the needs and preferences of producer and system stakeholders. With the help of proposed objective functions, producer stakeholders can enhance their appearance in the recommendation list. The system stakeholder can establish a balance between exposure of producer items and promotes items with low recommendation count. The proposed objective function for system stakeholders also provides a personalization facility for the producer by providing a unique user count to the producer. To observe and evaluate utility gain, we incorporated the objective function algorithm in the baseline algorithms. With the observed Pareto front and evaluation result, it is clear that the proposed stakeholder optimization-based recommendation model is capable of generating a trade-off recommendation solution. Also, it can fetch better utilities for producer and system stakeholders. This work presents the preliminary examination of stakeholder's objective optimization; this work could be extended by incorporating more stakeholders and designing utility functions that can also fetch economic utilities.

References

1. Aggarwal, C.C.: An introduction to recommender systems. In: Recommender Systems, pp. 1–28. Springer, Cham (2016). https://doi.org/10.1007/978-3-319-29659-3_1
2. Pujahari, A., Sisodia, D.S.: Aggregation of preference relations to enhance the ranking quality of collaborative filtering based group recommender system. Expert Syst. Appl. **156**, 113476 (2020). https://doi.org/10.1016/j.eswa.2020.113476
3. Burke, R., Abdollahpouri, H., Mobasher, B., Gupta, T.: Towards multi-stakeholder utility evaluation of recommender systems. In: CEUR Workshop Proc. 1618 (2016)
4. Abdollahpouri, H.: Incorporating system-level objectives into recommender systems. In: Web Conf. 2019 - Companion World Wide Web Conf. WWW 2019, vol. 2, pp. 2–6 (2019). https://doi.org/10.1145/3308560.3314201
5. Pujahari, A., Padmanabhan, V.: Group recommender systems: combining user-user and item-item collaborative filtering techniques. In: Proceedings of - 2015 14th International Conference Infornation Technology ICIT 2015, pp. 148–152 (2016). https://doi.org/10.1109/ICIT.2015.36.
6. Pujahari, A., Sisodia, D.S.: Modeling side information in preference relation based restricted Boltzmann machine for recommender systems. Inf. Sci. (Ny) **490**, 126–145 (2019). https://doi.org/10.1016/j.ins.2019.03.064
7. Pujahari, A., Sisodia, D.S.: Pair-wise preference relation based probabilistic matrix factorization for collaborative filtering in recommender system. Knowl.-Based Syst. **196**, 105798 (2020). https://doi.org/10.1016/j.knosys.2020.105798
8. Shrivastava, R., Sisodia, D.S.: Product recommendations using textual similarity based learning models. In: 2019 International Conference on Computer Communication and Informatics (ICCCI), pp. 1–7 (2019)
9. Pizzato, L., Rej, T., Chung, T., Koprinska, I., Kay, J.: RECON: a reciprocal recommender for online dating. In: RecSys 2010 – Proceedings of 4th ACM Conference on Recommender System, pp. 207–214 (2010). https://doi.org/10.1145/1864708.1864747
10. Xia, P., Liu, B., Sun, Y., Chen, C.: Reciprocal recommendation system for online dating. In: Proceedings of 2015 IEEE/ACM International Conference on Advances in Social Networks Analysis and Mining, ASONAM 2015, pp. 234–241 (2015). https://doi.org/10.1145/2808797.2809282
11. Mine, T., Kakuta, T., Ono, A.: Reciprocal recommendation for job matching with bidirectional feedback. In: Proceedings - 2nd IIAI International Conference on Advanced Applied Informatics, IIAI-AAI 2013, pp. 39–44 (2013). https://doi.org/10.1109/IIAI-AAI.2013.91
12. Evans, D.S., Schmalensee, R., Noel, M.D., Chang, H.H., Garcia-Swartz, D.D.: Platform economics: essays on multi-sided businesses. Compet. Policy Int. 459 (2011)
13. Zheng, Y., Ghane, N., Sabouri, M.: Personalized educational learning with multi-stakeholder opti-mizations. In: ACM UMAP 2019 Adjunct Publication of the 27th Conference on User Modeling, Adaptation and Personalization, pp. 283–289 (2019). https://doi.org/10.1145/3314183.3323843
14. Zheng, Y., Pu, A.: Utility-based multi-stakeholder recommendations by multi-objective optimization. In: Proceeding of - 2018 IEEE/WIC/ACM International Conference on Web Intelligence WI 2018, 128–135 (2019). https://doi.org/10.1109/WI.2018.00-98
15. Hernando, A., Bobadilla, J., Ortega, F.: A non negative matrix factorization for collaborative filtering recommender systems based on a Bayesian probabilistic model. Knowl.-Based Syst. **97**, 188–202 (2016). https://doi.org/10.1016/j.knosys.2015.12.018
16. Luo, X., Xia, Y., Zhu, Q.: Incremental collaborative filtering recommender based on regularized matrix factorization. Knowl.-Based Syst. **27**, 271–280 (2012). https://doi.org/10.1016/j.knosys.2011.09.006

17. Ge, M., Delgado-Battenfeld, C., Jannach, D.: Beyond accuracy: evaluating recommender systems by coverage and serendipity. In: RecSys 2010 - Proceedings of 4th ACM Conference Recommender on System, pp. 257–260 (2010). https://doi.org/10.1145/1864708.1864761
18. Vargas, S., Castells, P.: Rank and relevance in novelty and diversity metrics for recommender systems. In: RecSys'11 – Proceedings of 5th ACM Conference Recommender on System, pp. 109–116 (2011). https://doi.org/10.1145/2043932.2043955
19. Jain, A., Singh, P.K., Dhar, J.: Multi-objective item evaluation for diverse as well as novel item recommendations. Expert Syst. Appl. **139**, 112857 (2020). https://doi.org/10.1016/j.eswa.2019.112857
20. Cui, L., Ou, P., Fu, X., Wen, Z., Lu, N.: A novel multi-objective evolutionary algorithm for recommendation systems. J. Parallel Distrib. Comput. **103**, 53–63 (2017). https://doi.org/10.1016/j.jpdc.2016.10.014
21. Jungkyu, H.A.N., Yamana, H.: A survey on recommendation methods beyond accuracy. IEICE Trans. Inf. Syst. **100**, 2931–2944 (2017). https://doi.org/10.1587/transinf.2017EDR0003
22. Sarkar, J.L., Majumder, A., Panigrahi, C.R., Roy, S.: MULTITOUR: a multiple itinerary tourists recommendation engine. Electron. Commer. Res. Appl. **40**, 100943 (2020). https://doi.org/10.1016/j.elerap.2020.100943
23. De Pessemier, T., Dooms, S., Martens, L.: Design and evaluation of a group recommender system. In: RecSys'12 – Proceedings of 6th ACM Conference on Recommender System, pp. 225–228 (2012). https://doi.org/10.1145/2365952.2366000
24. Abdollahpouri, H., Essinger, S.: multiple stakeholders in music recommender systems, 2017–2018 (2017). http://arxiv.org/abs/1708.00120
25. Burke, R., Abdollahpouri, H.: Educational recommendation with multiple stakeholders. In: Institute of Electrical and Electronics Engineers (IEEE), pp. 62–63 (2017). https://doi.org/10.1109/wiw.2016.028
26. Zheng, Y.: Multi-stakeholder recommendations: case studies, methods and challenges. In: RecSys 2019 - 13th ACM Conference on Recommender System, pp. 578–579 (2019). https://doi.org/10.1145/3298689.3346951
27. Zheng, Y.: Multi-stakeholder recommendation: applications and challenges (2017). http://arxiv.org/abs/1707.08913
28. Zheng, Y.: Utility-based multi-criteria recommender systems, pp. 2529–2531 (2019)
29. Jannach, D., Adomavicius, G.: Price and profit awareness in recommender systems (2017). http://arxiv.org/abs/1707.08029
30. Sürer, Ö., Burke, R., Malthouse, E.C.: Multistakeholder recommendation with provider constraints. In: RecSys 2018 - 12th ACM Conference on Recommender System, pp. 54–62 (2018). https://doi.org/10.1145/3240323.3240350
31. Deb, K., Pratap, A., Agarwal, S., Meyarivan, T.: A fast and elitist multiobjective genetic algorithm: NSGA-II (2002)
32. Breese, J.S., Heckerman, D., Kadie, C.: Empirical analysis of predictive algorithms for collaborative filtering (2013). http://arxiv.org/abs/1301.7363
33. Sarwar, B., Karypis, G., Konstan, J., Riedl, J.: Item-based collaborative filtering recommendation algorithms. In: Proceedings of 10th International Conference World Wide Web, WWW 2001, pp. 285–295. Association for Computing Machinery, New York (2001). https://doi.org/10.1145/371920.372071
34. Koren, Y., Bell, R., Volinsky, C.: Matrix factorization techniques for recommender systems. Comput. Long. Beach. Calif. **42**, 30–37 (2009). https://doi.org/10.1109/MC.2009.263.

Evaluation of the Merits and Demerits Associated with a DIY Web-Based Platform for e-commerce Entrepreneurs

Adedeji Olushola Afolabi[1], Stephen Oluwatobi[1], Onyeka Emebo[1], Sanjay Misra[1](✉), and Lalit Garg[2]

[1] Covenant University, Ota, Nigeria
{adedeji.afolabi,stephen.oluwatobi,onyeka.emebo, sanjay.misra}@covenantuniversity.edu.ng
[2] University of Malta, Msida, Malta
lalit.garg@um.edu.mt

Abstract. Web programmers no longer have the sole right to develop websites for entrepreneurs. This is because web creation has been simplified by do-it-yourself (DIY) website builders with ready-made design templates. In light of this, the study evaluated the merits and demerits associated with a DIY web-based platform for e-commerce entrepreneurs. The study utilized a quantitative approach using a cross-sectional survey method on entrepreneurs trained to build MVP using a DIY website builder and other free tools. The data obtained were analyzed using statistical tools of mean scores, principal component analysis (PCA), and analysis of variance (ANOVA). The result showed that a DIY website builder's essential features were the availability of designer-made templates, e-commerce functionalities, and access to an endless stock of images. The perceived benefits associated with a DIY web-based platform were mainly its ease of use, design flexibility, and the requirement of zero coding skills. These perceived benefits are further categorized into ease benefits, design benefits, and storage benefits. The occurrence of poor or slow internet connection, inability to edit website via a mobile device, and fear of customer/payment information safety were the significant challenges associated with a DIY web design platform. In conclusion, the study showed a significant difference in the challenges associated with a DIY web design platform among the entrepreneurs in terms of paying more for each branded email accounts and freezing screen during web design. The study recommended that there is a need to increase awareness and training on a DIY web design capacity among entrepreneurs. In the age of technological advancement and already made templates for web designs, entrepreneurs can tap into the Internet's merits to increase their visibility and attract potential clients/customers.

Keywords: Do-it-yourself · e-commerce · Socio-economic · Startups · Web-based platform

L. Garg et al. (Eds.): ISMS 2020, LNNS 303, pp. 214–227, 2022.
https://doi.org/10.1007/978-3-030-86223-7_19

1 Introduction

The use of the Internet by different businesses have made the commerce world more competitive [1]. This aspect is reflective in other sectors of the economy, whereby the difference is created by those who engage actively in using the World Wide Web. Regardless of firms' size, diverse businesses across different sectors in the economy are integrating their services with an online community to engage half of the world's population connected to one form of social media. There is no doubt that they miss out on a large pool of potential customers for those still traditional and not engaging the online community. Although it is a new trend, it is widely spreading with companies committing large amounts of resources due to the commensurate benefits [2]. With many customers moving online to surf through pictures of products and sources for services, small and medium scale entrepreneurs are taking advantage of the medium to compete with larger firms and draw from online shoppers' pool [3, 4]. These internet-gains have been the proponent and driver to the increase in the number of e-commerce entrepreneurs worldwide. The e-commerce space offers businesses the opportunity to be strategic in their marketing and rendering of services without cognizance to their size or availability of an office environment. The small and large firms are brought to a leveled playing ground in the e-commerce environment. The process of e-commerce is explained by [5] and [6] as the act of displaying, advertising, and procurement of a firm's products, services, and information through different ICT (Information and Communication Technology) platforms which takes place between the firm and any third parties it deals with.

The rate of growth of the e-commerce site daily is astounding. Reports showed that over 20,000 new e-commerce stores are opened every week [7]. In creating a viable e-commerce platform for small and medium scale entrepreneurs, there are three main avenues. The first option is to work with a third-party e-commerce platform provider to help provide a platform to set up one's store, such as Amazon, eBay, and Google Shopping. The second avenue is to hire a web developer's professional services to develop the e-commerce site from start to finish. The third way is for e-commerce entrepreneurs to create their e-commerce store by themselves. This affords flexibility and control with little technical knowledge of web designs due to the availability of easy-to-use tools to deliver the entrepreneur's needs. The Do-It-Yourself (DIY) e-commerce web design can be tackled in two (2) ways. One option is to code the web-based e-commerce platform using HTML (interface design), PHP (Programming language), and MySQL (database management system, DBS) or using a web design software like Adobe Dreamweaver or Microsoft Expression Web. The second option this study discusses is a DIY that involves website builders such as WIX, Squarespace, Weebly, Site 123, GoDaddy, IM Creator, BigCommerce, and Shopify to help users use templates to build an e-commerce site within minutes. With the prospects of a DIY website, the study posits merits and demerits associated with a DIY web-based platform for e-commerce entrepreneurs. The task would be guided by the following objectives, which are to;

- Identify the features that e-commerce entrepreneurs appreciate in a DIY web-based platform.
- Evaluate the socio-economic benefits derived from a DIY web-based platform for e-commerce.

- Examine the inhibiting factors associated with the use of a DIY web-based platform.

This study showed that there are reasons to which e-commerce entrepreneurs use a do-it-yourself (DIY) web-based platform. The reasons can be due to the inherent features of the DIY web-based platform, the benefits of using the system or even the factors that makes it difficult to use the DIY web-based platform. By understanding these concepts in this study, it is envisaged that it will help developers of DIY web-based platform understand better their customers in the e-commerce sector. This should in turn increase the number of e-commerce entrepreneurs that use web-based technologies in attracting new clients and servicing old ones for the growth of their businesses.

2 Review of Related Literature

Entrepreneurs are highly crucial to the success of any economy globally. Therefore, entrepreneurs must succeed in a nation's economy due to their employment generation and contribution to the gross domestic product. The study by [8] stated that when considering economic growth and wealth creation among nations, e-commerce and entrepreneurship fields have significant impacts. Several studies have pointed to the success metrics that ensure that new ventures succeed. Such studies in [9] and [10] noted new ventures succeed solely based on the entrepreneur's characteristics. This is in line with findings from [2] that perceived usefulness and the perceived ease of use were the main determinants of adopting e-commerce among entrepreneurs. This means that to increase the adoption of e-commerce, there is a need to increase training among entrepreneurs and exposure [11]. Simultaneously, some other researchers believe that their success hinges on the industry's structure and the type of venture strategy implemented within the firm. This venture strategy will include the need and investment in the use of technology to assist the business. But why do entrepreneurs have the use of information technologies (IT) in new ventures? McQueen [12] opined that IT's inclusion in one's business might be due to experience in using the Internet or past business experiences that have engaged IT to use. The latter is corroborated in the study by [13], where 71% adopted IT in their new businesses due to previous experience in their business field. In all, entrepreneurs are always looking for new ways to make their businesses succeed and therefore are among the growing population associated with the use of the Internet.

E-commerce entrepreneurs conduct their commercial activities to provide goods and services with the aid of information and communication technologies connected to the Internet [5]. Therefore, been simplified e-commerce ventures are online businesses. They manage their orders, inventory, and business process transaction on different online web-based systems via the Internet. [14] noted that the enabling technology that makes e-commerce possible with high accuracy and efficiency is the Internet using various online platforms. These online platforms can be a company website, a third-party website, and other social media platforms such as Facebook, Instagram, Twitter, etc. With these IT platforms, e-commerce entrepreneurs can be more competitive, access new markets, and create new employment opportunities that help generate wealth and sustainable economic growth [15].

Furthermore, [16] stated the importance and impact of internet-based technologies on the success of businesses. Their study concluded that the Internet would only be useful to business managers that know how to use the innovative tool. This brings to fore the platforms that offer the do-it-yourself platforms for e-commerce entrepreneurs to maximize their reach to their present audience and a new audience.

3 Research Method

The study utilized a quantitative approach using a cross-sectional survey method to examine the socio-economic implications of a DIY web-based platform for e-commerce entrepreneurs to obtain the population to participate in the study. A cross-section of entrepreneurs was selected from a training sponsored by the World bank "starting a Startup class (HS3)" by Ventures Platform in partnership with Hebron Start-up Lab and CEDS Covenant University. The "How to start a startup series" dubbed HS3 was a one-week, interactive, classroom-style boot camp for entrepreneurs who want to build Africa's future through innovation. For the project's duration, the students are taught a methodological process to build a business from the idea stage. In training, entrepreneurs were taught how to create an MVP using WIX and other free tools. The Do-It-Yourself training on Wix.com was facilitated by Venture Africa and comprised of fifty-nine (59) trainees at the program. Various stages in creating a website for different entrepreneurs were treated during the program. Through the training, a google form was created and disseminated online to the entrepreneurs. The google form contained four (4) sections, including the personal details about the startup, the features in the DIY web design that best suits the startup, the challenges and benefits of using a DIY web design for entrepreneurs. From the responses, only thirty-two participants completed the online questionnaire, which represented a 54% survey response rate. The google form file was transferred in a.xls file to SPSS v. 21 for analysis. Statistical tools of frequencies, percentages, mean scores, principal component analysis (PCA), and analysis of variance (ANOVA) was performed on the data collected.

4 Results and Discussion

This section showed the analysis of data and discussion of the implications of findings. A summarized participants' information and analysis of the objectives in terms of features, benefits and challenges in using a DIY web design are presented.

4.1 Characteristics of Participants

In this study, thirty-two (32) participants from a DIY web design partook in the survey analysis out of the 59 trainees. It is essential to understand the entrepreneurs' characteristics of why they need the web design and what. Table 1 showed holistic features of the participants in terms of years in which the startup has been in existence, availability of office space, the number of employees in the startup, level of proficiency in using a DIY web design, usefulness of a DIY platform to the startup and the sector to which the startup provides service. In Table 1, most startups have only been founded within the last two years and do not have office spaces. The latter makes the need to have a website fundamental to the survival of the startups. Using a customized website would be essential for startups to reach a wider audience and potential customers. Table 1 showed that most of the entrepreneurs employed between 1 to 2 employees. This shows the crucial role of small businesses in creating employment. Therefore, this training would help increase the capacity of such enterprises. From the training, many

Table 1. Characteristics of participants in DIY web design

Characteristics of participants	Frequency	Percentage	Cumulative percentage
Years of startup existence			
0–2 years	20	62.5	62.5
3–5 years	12	37.5	100.0
Availability of office space			
No	31	96.9	96.9
Yes	1	3.1	100.0
Number of employees			
1 employee	13	40.6	40.6
2 employees	14	43.8	84.4
3 employees	5	15.6	100.0
Level of proficiency in using a DIY web design			
Basic knowledge	2	6.3	6.3
Limited experience	5	15.6	21.9
Intermediate	13	40.6	62.5
Advanced	12	37.5	100.0
The usefulness of a DIY platform to your business			
Low usefulness	1	3.1	3.1
Neutral	3	9.4	12.5
Moderately useful	15	46.9	59.4
Very useful	13	40.6	100.0
Startup sector			
Commerce (C)	12	37.5	37.5
ICT	11	34.4	71.9
Fashion (F)	4	12.5	84.4
Entertainment (ENT)	2	6.3	90.6
Education (EDU)	1	3.1	93.8
Others	2	6.3	100.0

of the entrepreneurs had become familiar with the DIY web design platform and found it useful in growing their businesses' visibility. The startups cut through several sections of the economy from commerce, ICT, fashion, entertainment, and education. These are areas where electronic assistance can increase the reach of startups.

4.2 Features of a DIY Web Design that Aligns with the Startups

There are different features of a DIY web design interface. These features appeal to other users in diverse ways. The study sought to show the best features that align with each startup. The study used a 5-point Likert scale of importance to elicit information on a DIY web platform's best features. Table 2 showed the features that the entrepreneurs found necessary in the operation of their businesses. Using compare means, Table 2 broke down the elements that appeal to each startup. From Table 2, the availability of designer-made templates, e-commerce functionalities (display of goods and services, delivery services, pricing system, generation of coupons/discount codes and shopping cart), and access to an endless stock of images was the overall most essential features in a do-it-yourself web design. These findings are crucial to web-designers as it shows the aspects entrepreneurs are concerned about in their websites' functionalities. It is not surprising that new-age entrepreneurs are worried about time and quality. With the feature of templates that entrepreneurs can select from, DIY websites can be encouraged. For example, one of the DIY website builder boast of having over 510 beautiful templates that different entrepreneurs from diverse sectors can select for their DIY website. The templates are categorized so that entrepreneurs can easily choose according to their business characteristics. All these things are made possible, even without any form of coding skills by the entrepreneur. The functionalities of the DIY websites must engender the attributes of the entrepreneurs. The functions should be easy to incorporate on the DIY websites. This will enhance customization using the drag-and-drop tool of DIY website builders. As e-commerce entrepreneurs are mostly about displaying goods and services, the availability of an endless amount of pictures to showcase their prowess is essential to their businesses.

4.3 Benefits Associated with a DIY Web Design Platform

The future of a do-it-yourself (DIY) web design platform can be attributed to the benefits that can be derived from it. Different benefits were harvested from feedback in the training and reviews from various sites of DIY web platforms. Table 3 showed the benefits that can be derived from using a DIY web platform by the entrepreneurs. In Table 3, the mean score analysis showed that the perceived benefits of a DIY web-based platform were mainly its ease of use, design flexibility, and zero coding skills needed to design the DIY web platform. Several theories on the use of new technology, such as Unified Theory of Acceptance and Use of Technology (UTAUT), Technology Acceptance Model (TAM), and Rogers' Diffusion of Innovation (DOI) Theory, emphasized the need for a technology to exhibit ease of use in its performance. This would help determine the speed at which prospective users would adopt the system. The research by [17] opined that one of the most critical factors entrepreneurs try to use technology to satisfy potential customers is the ease of use. To evaluate most systems' services, including web-based

Table 2. Features of a DIY web design

Features of a DIY web design	Startup sector							
	C	ICT	F	ENT	EDU	Others	Total	
	MS	MS	MS	MS	MS	MS	MS	RI
Availability of designer-made template	4.83	4.82	4.75	4.50	5.00	4.00	4.75	1st
Mobile editor interface	4.58	4.18	4.75	4.00	4.00	3.50	4.34	5th
Availability of one URL for website and mobile site	4.00	3.91	4.25	3.50	4.00	3.00	3.91	12th
Customized domain names	4.00	3.64	4.00	3.50	4.00	3.50	3.81	15th
Tools for search engine optimization (SEO)	4.58	3.91	4.00	4.00	5.00	4.00	4.22	6th
Free and reliable hosting	4.08	4.09	4.00	4.00	4.00	3.50	4.03	9th
Availability of support systems	3.75	3.91	4.00	4.00	4.00	3.50	3.84	14th
Ability to blog on the site	4.08	3.64	3.75	3.50	3.00	3.00	3.75	18th
Availability of social media buttons	4.75	4.18	4.25	4.50	4.00	4.00	4.41	3rd
e-commerce functionalities (display of goods and services, delivery services, pricing system, generation of coupons/discount codes and shopping cart)	4.75	4.09	5.00	3.50	4.00	4.50	4.44	2nd
Newsletter creator	3.92	3.64	3.75	3.50	3.00	4.00	3.75	18th
Membership creator	3.75	3.82	3.75	3.50	5.00	4.00	3.81	15th
Availability of plug-in apps	3.92	4.09	4.00	3.50	5.00	4.00	4.00	11th
Integration of Google maps	4.25	4.18	4.00	4.50	5.00	4.00	4.22	6th
Access to endless stock of images	4.58	4.18	4.75	4.00	4.00	4.50	4.41	3rd
An easy to use search bar	4.00	3.91	3.75	3.00	5.00	4.00	3.91	12th
Availability of email marketing services	4.25	4.00	3.75	3.00	5.00	4.00	4.03	9th
Multiple editors	4.17	4.27	3.75	3.50	4.00	4.50	4.13	8th
Security of website	3.75	3.91	3.75	3.00	5.00	4.00	3.81	15th

**MS = Mean Score; RI = Ranking Index*

systems, many researchers use the factor of ease of use [18]. According to [19], this ease of use involves a minimum requirement to access and move through the online system. This would help reduce customer frustration and increase client satisfaction [20].

To further understand the benefits variables identified in this study, and the similarities within the components, a principal component analysis (PCA) was performed. The PCA performed on a DIY web platform's benefits showed a 3-component solution of 33.3%, 16.4%, and 11.2% variance in the eigenvalues. A total of 61% variance was explained in

Table 3. Benefits associated with a DIY web design platform

Benefits	Mean score	Std. deviation	Ranking index
It is easy to use	4.81	0.397	1st
Design flexibility	4.59	0.499	2nd
Zero coding skills or design skills	4.53	0.621	3rd
Responsive themes	4.41	0.499	4th
User-friendly interface	4.38	0.492	5th
Increases innovation	4.31	0.693	6th
Availability of varieties in themes, pictures, and templates	4.31	0.535	6th
Availability of storage space for online content	4.22	0.553	8th
Fast speed in creating a website	4.19	0.535	9th
Increased control over the website	4.19	0.592	9th
Adequate content management	4.19	0.644	9th
Ease of duplicating pages	4.16	0.628	12th
No need to hire a professional web designer	4.16	0.723	12th
Changes can be achieved quickly without contacting a web designer	4.13	0.609	14th
It requires little or no budget to build	4.00	0.672	15th
Availability of online help support system	3.97	0.595	16th
Affordable pricing	3.84	0.723	17th

the 3-component model summary and is deemed adequate for other component loadings. Table 4 showed the principal component analysis of the benefits associated with a DIY web platform. Table 4 shows three components: Ease the benefits of DIY systems, the Design benefits, and Storage benefits. The ease benefits associated with the use of DIY web design platforms are mainly attributed to the help support that the companies give when an individual is stuck in the design process, the speed of design, the zero coding skill required, increase in innovation, availability of varieties in themes, pictures and templates, the ease of duplicating the web pages and the ease of using the system. These characteristics endear entrepreneurs to take up the use of DIY web platforms. According to [21], ease of use in measuring the e-quality of e-commerce systems by entrepreneurs involves the ease at which users can learn the new system, ease to remember the layout, ability to know what to do with little support, low physical effort, and minimum skill required. Another significant benefit is the design benefit that DIY web design platforms give to users. Due to the detailed design by the DIY web design platforms and users' consideration, there is design flexibility for different users and a user-friendly interface. Users can select designs that fit into their service or sector and customize the outlook for their customers. Lastly, very crucial to entrepreneurs is the need to manage their content from time to time. DIY web platforms offer Users adequate storage space for their online content in pictures, transactions, and inventory. These benefits associated with DIY web platforms can help developers increase users' number to their medium by marketing these features to potential clients.

Table 4. PCA for benefits associated with DIY web design platforms

Components	Dimension		
	1	2	3
Availability of online help support system	0.809	−0.017	0.451
Fast speed in creating a website	0.776	0.012	−0.142
Zero coding skills or design skills	0.765	−0.443	−0.278
Increases innovation	0.734	−0.470	−0.254
Availability of varieties in themes, pictures, and templates	0.731	−0.273	−0.138
Increased control over the website	0.714	−0.243	−0.370
Ease of duplicating pages	0.688	0.089	−0.085
It requires little or no budget to build	0.615	0.124	0.458
No need to hire a professional web designer	0.589	0.532	−0.218
Design flexibility	−0.031	0.758	−0.162
User friendly interface	0.219	0.707	−0.324
It is easy to use	0.054	0.507	−0.269
Availability of storage space for online content	0.387	0.199	0.595

4.4 Challenges Associated with a Do-It-Yourself Web-design

In a DIY web design platform, some challenges could be experienced in trying to use the system. Users of the platform were asked about challenges associated with the use of DIY web platforms. The challenges were obtained from feedback on DIY web training and reviews on DIY web design platforms. Table 5 showed the mean score analysis of the difficulties associated with a DIY web design platform. In Table 5, the occurrence of poor or slow internet connection, inability to edit website via a mobile device, and fear of customer/payment information were the significant challenges associated with a DIY web design platform. Since DIY web design platforms are web-based, there is the need to provide a fast internet connection to users when designing. The issue of slow or poor internet connection can discourage users from proceeding with DIY web design platforms. This is a crucial challenge in sub-Saharan Africa; the Internet is slow and accessible at exorbitant prices. Besides, using DIY web design platforms, users encountered difficulties in using mobile devices to edit their websites. Most entrepreneurs are most times on the move with their mobile devices making different transactions.

Table 5. Challenges associated with a DIY web design platform

Challenges	Mean score	Std. deviation	Ranking index
The occurrence of poor or slow internet connection	4.41	0.911	1st
Inability to edit website via a mobile device	4.28	0.772	2nd
Fear of the safety of customer/payment information	3.91	0.818	3rd
Difficulty in moving the website to another hosting site	3.88	0.660	4th
No live chat support option on the website builder	3.84	0.884	5th
Low flexibility in changing themes after completion of the website	3.84	0.767	5th
Non-availability of hardware such as laptop to design it	3.84	0.723	5th
Not fully mobile responsive	3.81	0.780	8th
High long term pricing of the platform	3.78	0.870	9th
Fear of online content information being stolen	3.75	0.880	10th
Freezing of screen during the design	3.66	0.865	11th
Poor loading time	3.66	0.745	11th
Paying more for each branded email accounts	3.63	0.751	13th
Constant ads on the platform	3.50	0.916	14th
Power supply issues	3.22	0.975	15th

Another major challenge is the fear that customers' payment information may not be secured due to the DIY web platforms' third-party ownership.

To validate the challenges associated with using a DIY web design platform, the study tested a hypothesis on whether there was a significant difference in the challenges associated with a DIY web design platform among the entrepreneurs. Using an analysis of variance (ANOVA) test, the study analyzed the difficulties of using a DIY system such as website design. All variables that had significant values less than or equal to 0.05 were termed significant. Table 6 shows that paying more for each branded email accounts and freezing screen during design were significant challenges each of the entrepreneurs faced in using a DIY web design platform. Most DIY website builders do not come with email plans, and therefore they are handled as a separate service to the user, which would require more cost. While using a DIY website builder, the freezing of the screen is due to a buildup of information on the individual's computer, resulting in a standstill of the DIY website or sometimes crash. This problem has a step-by-step solution available on several DIY websites, and support systems can resolve such issues. However, DIY website developers need to fix such errors to reduce the frustration from loss of time in using the website builders.

Table 6. ANOVA test on the challenges associated with a DIY web design platform

Challenges in a DIY web design platform		Sum of squares	Df	Mean square	F	Sig.
Occurrence of poor or slow internet connection	Between groups	6.173	5	1.235	1.642	0.184
	Within groups	19.545	26	0.752		
	Total	25.719	31			
Non-availability of hardware such as laptop to design it	Between groups	5.143	5	1.029	2.415	0.063
	Within groups	11.076	26	0.426		
	Total	16.219	31			
Power supply issues	Between groups	7.120	5	1.424	1.657	0.180
	Within groups	22.348	26	0.860		
	Total	29.469	31			
Fear of online content information being stolen	Between groups	4.697	5	0.939	1.265	0.309
	Within groups	19.303	26	0.742		
	Total	24.000	31			
Fear of the safety of customer/payment information	Between groups	1.969	5	0.394	0.546	0.740
	Within groups	18.750	26	0.721		
	Total	20.719	31			
High long term pricing of the platform	Between groups	4.052	5	0.810	1.085	0.392
	Within groups	19.417	26	0.747		
	Total	23.469	31			
Constant ads on the platform	Between groups	7.197	5	1.439	1.990	0.114
	Within groups	18.803	26	0.723		
	Total	26.000	31			
Low flexibility in changing themes after completion of the website	Between groups	2.310	5	0.462	0.755	0.590
	Within groups	15.909	26	0.612		
	Total	18.219	31			
Not fully mobile responsive	Between groups	2.966	5	0.593	0.969	0.455
	Within groups	15.909	26	0.612		
	Total	18.875	31			

(continued)

Table 6. (*continued*)

Challenges in a DIY web design platform		Sum of squares	Df	Mean square	F	Sig.
No live chat support option on the website builder	Between groups	7.552	5	1.510	2.356	0.069
	Within groups	16.667	26	0.641		
	Total	24.219	31			
Paying more for each branded email accounts	Between groups	7.197	5	1.439	3.632	0.013
	Within groups	10.303	26	0.396		
	Total	17.500	31			
Poor loading time	Between groups	4.916	5	0.983	2.078	0.101
	Within groups	12.303	26	0.473		
	Total	17.219	31			
Freezing of screen during the design	Between groups	10.393	5	2.079	4.214	0.006
	Within groups	12.826	26	0.493		
	Total	23.219	31			
Inability to edit website via a mobile device	Between groups	3.923	5	0.785	1.403	0.256
	Within groups	14.545	26	0.559		
	Total	18.469	31			
Difficulty in moving the website to another hosting site	Between groups	3.674	5	0.735	1.944	0.121
	Within groups	9.826	26	0.378		
	Total	13.500	31			

5 Conclusion and Recommendation

Web programmers no longer have the sole right to develop websites for entrepreneurs. This is because DIY website builders have harnessed the use of already designed templates for different service sectors, thereby simplifying website creation. The study evaluated the merits and demerits of a do-it-yourself (DIY) web-based platform for e-commerce entrepreneurs. Entrepreneurs that use DIY web design platforms are mostly interested in features that ensure availability of designer-made templates, availability of e-commerce functionalities (display of goods and services, delivery services, pricing system, generation of coupons/discount codes, and shopping cart), and access to an endless stock of images. The perceived benefits of a DIY web-based platform were mainly its ease of use, design flexibility, and zero coding skills needed to design the DIY web platform. These perceived benefits are further categorized into ease benefits, design benefits, and storage benefits. The occurrence of poor or slow internet connection, inability to edit website via a mobile device, and fear of customer/payment information safety were the significant challenges associated with a DIY web design platform. In conclusion, the study showed a significant difference in the challenges associated with a DIY web design platform among the entrepreneurs in terms of paying more for each branded email accounts and freezing screen during web design. The following recommendations are put forward in this study;

- There is a need to increase awareness and training on a DIY web design capacity among entrepreneurs. In the age of technological advancement and already made templates for web designs, entrepreneurs can tap into the Internet's merits to increase their visibility and attract potential clients/customers.
- For small firms that may want to scale and migrate easily, there is a need for website builders to aid the transferability of content to a different hosting site. Also, as the firms begin to scale, the need to increase the speed of loading the sites may be an issue due to more visitors.
- As small-scale entrepreneurs build their client base, there is the constant fear of losing client information, including payment details, and so on. Website builders need to continuously affirm the privacy of businesses that are hosted on their shared hosting platforms.
- Website building sites need to provide an adequate online help support system to deal with several challenges of users such as freezing of site or total outage of website.

Acknowledgment. The authors appreciate the sponsorship from Covenant University through its Centre for Research, Innovation and Discovery, Covenant University, Ota Nigeria.

References

1. Afolabi, A.O., Ojelabi, R.A., Fagbenle, O.I., Mosaku, T.O.: The economics of cloud-based computing technologies in construction project delivery. Int. J. Civil Eng. Technol. (IJCIET) **8**(12), 232–242 (2017)
2. Lane, P.P.M., Wafa, S.A., Hassan, R.A., Belkhamza, Z.: Perceived usefulness and perceived ease of use of e-commerce adoption among entrepreneurs in Sabah. Kuwait Chapter Arab. J. Bus. Manage. Rev. **3**(9), 94–103 (2014)
3. Afolabi, A., Ojelabi, R., Oyeyipo, O.: An analysis of social media marketing of indigenous construction firms in Nigeria: a tool for sustainable growth. Int. J. Constr. Supply Chain Manage. **8**(2), 60–72 (2018)
4. Ojelabi, R.A., Oyeyipo, O.O., Afolabi, A.O., Amusan, L.M.: Presence of client social relationship management within the Nigeria construction industry. Buildings **8**(4), 60–75 (2018)
5. Kabugumila, M.S., Lushakuzi, S., Mtui, J.E.: e-Commerce: an overview of adoption and its effective implementation. Int. J. Bus. Soc. Sci. **7**(4), 243–252 (2016)
6. Afolabi, A., Ibem, E., Aduwo, E., Tunji-Olayeni, P., Oluwunmi, O.: Critical success factors (CSFs) for e-Procurement adoption in the Nigerian construction industry. Buildings **9**(47), 1–18 (2019)
7. Wong, C.K.: An untapped e-commerce opportunity. [Online] Retrieved December 1, 2010.http://www.zippycart.com/ecommerce-news/1186-malaysia-an-untapped-ecommerce-opportunity.html (2010)
8. Acs, Z., Arenius, J.P., Hay, M., Minniti, M.: Global entrepreneurship monitor: 2004 executive report, Boston, MA, Babson College and UK: London Business School (2004)

9. Busenitz, L.W., Lau, C.-M.: A cross-cultural cognitive model of new venture creation. Entrepreneurship Theor. Pract. **20**(4), 25–39 (1996)
10. Blin, F.: Breaking the growth barrier. Director **51**(9), 69 (1998)
11. Belkhamza, Z.N., Wafa, S.A.: The effect of perceived risk on the intention to use e-commerce: the case of Algeria. J. Internet Bank. Commer. **14**(1), 1–10 (2009)
12. McQueen, R.J.: e-entrepreneurship: Proposed use of the Internet in a business plan competition in New Zealand, Hamilton, New Zealand. Department of Management Systems, University of Waikato (2014)
13. Mazzarol, T., Volery, T., Doss, N., Thein, V.: Factors influencing small business startups. Int. J. Entrepreneurial Behav. Res. **5**(2), 48–63 (1999)
14. Applegate, L.M., Austin, R.D., McFarlan, F.W.: Creating Business Advantage in the Information Age. McGraw-Hill, New York (2002)
15. Sebora, T.C., Lee, S.M., Sukasame, N.: Critical success factors for e-commerce entrepreneurship: an empirical study of Thailand. Small Bus. Econ. **32**, 303–316 (2009)
16. Barhatov, V., Campa, A., Pletnev, D.: The impact of internet-technologies development on small business success in Russia. Proc. Soc. Behav. Sci. **238**, 552–561 (2018)
17. Korper, S., Ellis, J.: The e-commerce Book: Building the e-empire, 2nd edn. Academic Press, New York (1999)
18. Lockett, A., Littler, D.: The adoption of direct banking services. J. Mark. Manage. **13**, 791–811 (1997)
19. Zeithaml, V.A., Parasuraman, A., Malthotra, A.: A conceptual framework for understanding e-service quality: implications for future research and managerial practice. Working Paper, Report Nr. 00–115, Cambridge, MA: Marking Science Institute (2000)
20. Greenstein, M., Vasarhelyi, M.: Electronic Commerce: Security, Risk Management, and Control, 2nd edn. McGraw-Hill, New York (2002)
21. Betson, J.E.: Self-service encounter: an exploratory study. J. Retail. **61**(3), 46–76 (1985)

Check for updates

Link Prediction Based on Spatio-Temporal Networks

Kelly Steer and Joseph G. Vella(✉)

Faculty of Information and Communication Technology, University of Malta, Msida, Malta
{kelly.steer.12,joseph.g.vella}@um.edu.mt

Abstract. In a network composed of complex entities and relationships, its topology offers multiple patterns that could indicate different properties such as importance, rank, and category. These properties indicate new trends through their deeper analysis. A selection of social activities and interactions are not only dynamic, but their strength and reach evolve over time and locality. Anticipating the likelihood of future social interactions is similar to the Link Prediction Problem. This paper describes how a social network was built from a snapshot of a spatio-temporal dataset which includes user identifier, geo-coded location, and time event as attributes for people checking-in certain localities. This social network is then used as a basis to predict the likelihood of two persons checking-in at the same place over a comparable time interval. A set of features is used to hold scores indicating the similarities of pairs of nodes. One of the prediction features employed in this study is a time-based weight which describes the activeness of the network nodes based on how recent their adjacent interactions are. A supervised binary classification technique is used with these features on part of the dataset to segment results based on whether a link is formulated in the future. The model created is then used on a distinct test set to generate predictions. The results of this empirical study, yielded an overall accuracy of more than 90%. Other performance measures such as precision, positive prediction rate, and negative prediction rate are also used to aid the model's evaluation.

Keywords: Link prediction problem · Spatio-temporal data · Social networks

1 Introduction

Graph mining is a popular area studied by various researchers due to the rise in popularity of social networks where researchers strive to understand the topology of such networks to enable deeper knowledge extraction and thus provide better analytics.

A popular approach is the Link Prediction Problem (LPP) used to predict the likelihood of a future association between two nodes, given that there is no direct association between them in the current state of a graph [1]. Predicting these associations is beneficial in various applications which are interested in uncovering new trends, providing better recommendations, and discovering future relations.

The connectedness of nodes in networks is known to reveal certain patterns which in turn are used as a basis for future interaction prediction. Popular metrics from graph

L. Garg et al. (Eds.): ISMS 2020, LNNS 303, pp. 228–239, 2022.
https://doi.org/10.1007/978-3-030-86223-7_20

theory are adopted to expose features related to the graph structure. Various studies, such as [2–5], built models based on these features to predict whether a link is likely in the future. In addition to graph features, one could also use other attributes related to properties found in the data itself. Examples of these properties include demographics, time-based features, and locations.

This study uses a spatio-temporal dataset to build a social network describing people being at the same location during the same time interval. Four graph features, a time-based feature, and a Boolean value defining whether two users are online friends, form the basis of the classification computation. This classification is used to build a model which is used to predict future associations. In this case, an association is two users checking in at the same place over a specific time frame.

The results obtained are compared with the actual data to depict whether the predicted values are correct. Various measures such as accuracy, precision, and recall are used to evaluate the performance of the results.

2 Literature Review

2.1 Background Theory

In a graph *G*, two finite sets exist; one representing the vertices in the graph and the other represents edges connecting these vertices. In a typical social graph, the vertices represent different users, whereas the edges represent activities and interactions between users.

The majority of the analysis done on networks are based on these edges as they convey patterns useful in various fields. In domains where such links evolve with time, as in social networks, the main aim is to predict the occurrence of a link between two entities in the future, based on their attributes and their connections [6]. This is the LPP.

As previously outlined, metrics used in graph theory and social network analysis are widely used to convey similarities or proximity between pairs of nodes. Having high similarities could help to influence the probability of whether a particular link between a pair of nodes will occur. These metrics are commonly referred to as baseline metrics.

A number of studies focusing on the LPP used these metrics to assign weights to potential links such as [2–4], whereas others like [5, 11, 12], proposed improvements to these metrics based on their application. These weights correspond to specific pairs of nodes, represented as *Score(x,y)*, which denote the probability that a link between nodes *x* and *y* is created in the future. A ranking is thus produced, consisting of a list of scores in descending order.

Metrics Based on Node Neighbourhoods. Considering two nodes *x* and *y* in a graph let $\Gamma(x)$ be the set of neighbours of *x* in *G* and $\Gamma(y)$ the set of neighbours of *y* likewise in *G*. These metrics are based on the notion that the greater the overlap between $\Gamma(x)$ and $\Gamma(y)$, the higher the probability that a connection between *x* and *y* is created in *G* [2]. The following are common node based metrics for a graph *G*:

Sum of Neighbours. When adding the number of connections a specific node has the sum of neighbours is obtained. This is finding the degree of a node [1].

Common Neighbours. This calculation enumerates the number of common neighbours of two nodes. This is calculated by comparing the neighbours of a particular node with the neighbours of another node, and is calculated by the following formula [2–4]:

$$Score\,(x, y) := |(x) \cap (y)| \tag{1}$$

Jaccard's Coefficient. This computation simply divides the number of common neighbours by the total number of neighbours [7] and is expressed by the following [2]:

$$Score\,(x, y) := \frac{|\Gamma(x) \cap \Gamma(y)|}{|\Gamma(x) \cup \Gamma(y)|} \tag{2}$$

Adamic/Adar. Similar to the previous one, this metric is used to calculate the similarity between two nodes but instead of counting only the common neighbours, this is also based on their degrees [2]. In the formula below, z is considered to be the degree or a feature shared by x and y [2, 3]:

$$Score\,(x, y) := \sum\nolimits_z \frac{1}{\log(frequency(z))} \tag{3}$$

Metrics Based on Node Paths. This sub-section outlines a number of metrics which are based on the paths formed from one node to another by following connecting edges. The notion used here suggests that nodes which are closer together in a network are more likely to be connected in the future. The following are common path based metrics:

Shortest Distance. The shortest distance refers to the shortest number of hops required to get to a specific node. Additionally, the shortest distance could be weighted; this means that each edge contains a weight which is summed up to calculate the distance between two nodes.

Katz. This measure deals with the summation over all possible paths from x to y with exponential damping by length to weight short paths more heavily [2]. It is calculated by the following equation where $|\,paths\,(l)_{x,y}|$ is the set of all *length-l* paths from x to y [2–4]:

$$Score(x, y) := \sum\nolimits_{l-1} \beta \cdot |paths(l)_{x,y}| \tag{4}$$

Hitting Time. The hitting time $H_{x,y}$ from x to y is the expected number of steps required for a random walk starting at x to reach y [2]. The summation of the number of steps from x to y, and the number of steps from y to x is known as the commute time i.e. the summation of $H_{x,y}$ and $H_{y,x}$ gives the commute time.

PageRank. This measure works by counting the number of links a node has. Links can be incoming or outgoing. This measure also considers the quality of each link to roughly determine how important that node is. The more 'important' a node is, the more likely to create links to other nodes.

SimRank. This is a similarity measure based on the notion that two nodes are considered to be similar if they are referenced (or connected) by similar nodes [2]. This metric can also considers the direction (i.e. incoming or outgoing) of the link to evaluate a score.

These metrics provide statistical measures which are also sometimes used as features in supervised methods. Supervised learning is a task done in machine learning which uses labelled data to infer results. These methods are solely classification techniques as models are trained to learn how data could be classified to different classes (or labels). In fact, the LPP can be posed as a binary classification problem as it is used to distinguish positive and negative links.

In addition to graph features, other properties related to the data can be used to train the model in the classifier. Several classifiers exist to handle such predictions. Some of these include support vector machines (SVM), decision trees, and k-nearest neighbours.

Achieving correct predictions is not a trivial task due to multiple challenges posed by graphs. In reality these networks tend to be very sparse; i.e. the network consists of nodes which are connected to a very small fraction of all the nodes in that network [8]. This is challenging as it makes it more difficult to predict future associations since the calculated probability would not be as accurate as if the network was densely connected. Other challenges related to graph structures include size and graph features. The features chosen act as the base of the classifications and it is therefore essential to choose the most suitable features for a network. Graphs, such as social graphs, tend to grow at a fast pace with nodes and edges that are constantly being added. As graphs grow and change with time, they are capturing more complex social interactions and hence become more difficult to analyse.

The high dynamism portrayed in social graphs beckons the need of a new dimension which might help improve the accuracy of the results obtained in predictive models. Many believe that by considering time as a dimension the results achieved are improved significantly [9, 10].

2.2 Related Works

The outlined baseline metrics have been evaluated by a number of researchers, where they compared the precision achieved by each of them. Studies, for example in [2–4], use these metrics to predict future interactions in the datasets they used.

Each study uses the same approach where they calculated scores using various metrics. Each of them returned a ranked list consisting of predictions in descending order of confidence. The top predictions were then compared with the actual connections. The intersection between these two symbolized the accuracy obtained by each metric. The Katz metric was reported to perform really well in all the studies, and other simple metrics such as common neighbours and Jaccard's Coefficient also gave good results.

These baseline metrics are used by a number of studies to compute link predictions. Other researchers tried to alter these methods to better fit the application and domain in question. The study in [11], proposed methods based on the Katz metric and common neighbours. They compared results from the traditional methods with the proposed methods, where the latter performed better. Another study [12], proposed a variant of the common neighbours method. Their method suggests that the connections between uncommon neighbours are also considered as they play a crucial role which might improve the prediction of new links and also missing links; i.e. interactions that exists but not encoded in the graph. The proposed method was tested on four different real world networks and the results yielded an area under curve (AUC) value higher than

0.5. The study in [5] looks at the link prediction problem under the uncertain framework where edges are given a score as to whether they actually exist or not. They use neighbourhood information to assist in predicting new links and managed to perform better when compared to traditional methods.

In addition to graph features, other studies choose features from the data itself. The study in [13], used various node features for the link prediction on location based social networks. Since this quoted study focused on spatial analysis, the features chosen were based on geo-locations. They suggested that users who check-in the same places are more likely to make contact. Additionally, the geographic distance between two users might propose whether they will interact at a future point. Using these features allowed them to reduce the prediction possibilities by about 15 times by focusing on certain interactions. The model created managed to discover a good percentage of all new links.

As mentioned already, these graph based features together with other features are sometimes used in supervised techniques. The study in [1] evaluated various classifiers on the LPP to see which of these obtained the highest accuracy. They used features such as sum of neighbours and shortest distance to build classification models using several classification algorithms such as SVM, Decision Trees, Multilayer Perceptron, and K-Nearest Neighbours. The experiment was run on two separate datasets and which both pointed out SVM with a radial basis function (RBF) kernel as being the best classifier used in terms of accuracy. The study in [14] proposed a measure based on path depth and nodes degree which was used with various baseline features in supervised learning algorithms, such as random forest and k-nearest neighbours. Adding the proposed measure as a feature helped them obtain a higher accuracy when predicting new links.

Furthermore various researchers analysed the notion that by considering time as a dimension, the predictions would be improved. The study in [9] was one of the first studies which looked at this notion. They evaluated how static metrics perform, when compared to other techniques which used temporal attributes in their analysis. In fact their results show that temporal metrics are an extremely valuable new contribution to link prediction, and should be used in future applications. Another study [15], calculated the performance of several baseline metrics and also evaluated the performance of five new scores which take time into consideration when predicting new links. These techniques were tested on two social networks and again confirmed that considering temporal data found in links is beneficial as it yields improved link prediction.

Other studies [16, 17], used approaches based on time-series. They created a time-series based on score obtained from baseline metrics. A forecasting model such as the weighted moving average was used to calculate new scores and in turn feed in supervised techniques to classify positive and negative links. Both studies asserted significant improvements over baseline metrics and other supervised methods which does not consider the time aspect. One could find various studies which analysed temporal behaviour on the algorithms used for the LPP. Most of these agree that this consideration is beneficial as the results obtained are improved. This study uses a time-based and other features to build a model which predicts whether two people are likely to check-in at the same place over the same time interval. The following section outlines the methodology used to obtain such predictions.

3 Methodology

3.1 Data

The dataset used in this study is called Brightkite and is available from [18]. It is comprised of data describing users checking-in at a location at a specific time. The attributes include a user ID, check-in time, longitude, latitude, and location ID. Since this study is focused on social networks this data had to be transformed so that it is in a graph format i.e. consisting of nodes and edges.

A subset of this dataset, filtered data from March 2008 until September 2008, was loaded in a graph database (see Table 1). Nodes signifying persons and locations were used to identify each person and location found in the dataset. Links, represented by relationship *chekedin*, between persons and locations were created to represent people checking-in locations. This data was then used to build a social network connecting people who checked-in at the same location over the same period of time.

All pairwise combinations were matched to check whether the two persons checked-in at the same location on the same day. More specifically, if user a and user b were in location z on the same day, a link is created between user a and user b. Each link (represented by relationship *near*) includes a property describing the day when they were both in the same location. More than one link could exist between two users as they might have been at the same location on the same day multiple times.

The dataset also included information about these online users. The data shows whether two users are online friends. This was also loaded in the graph database by matching pairs and checking whether they are online friends. If they are, a new link (encoded by relationship *friends*) is created between the two nodes.

The main experiment uses the social network describing persons and edges with the near relationship. This is explained in more detail in the following sub-section.

Table 1. Data volumes and statistics

Type	Label	Total
Node	Person	1387
Node	Location	24425
Relationship	Checked in	138796
Relationship	Near	13554
Relationship	Friends	18130

3.2 Experiment

The main experiment undertaken in this study uses a supervised approach to build a classification model using labelled data. The classification model is built based on six features chosen to represent the data and the network topology. The first feature is based on time factors. This was included in this experiment since previous studies have empirically concluded that considering such attributes help achieve improved predictions.

The time-based feature used here revolves around the activeness of nodes. This feature is therefore a weight representing how active a node is. Nodes having higher weights are considered to be key influencers in a network. Such nodes are more likely to form new links and hence predictions involving them are likely to be positive.

Considering the number of interactions a node has indicates the activity of the node. This however is a static measure and does not reflect when these links were created. To put this into perspective, imagine there is a network of people writing papers together, a particular author has written a good number of papers, and hence it is perceived as being an influential node in the network since its degree would be high. In a static link prediction method, this person is one of the focus points due to his activity and connectedness in the network. Therefore, any predictions involving him with other active nodes are more positively skewed, which means that new links with this node are more likely to be predicted. However, this does not consider when the node was active; what if the last paper that the author wrote was a couple of years ago? Should he still be considered as an influential author?

Due to these possibilities, it is useful to consider when each interaction has happened. Consequently, each interaction is represented as a time event which describes when the two users were at the same place on the same interval (e.g. a day). The objective of this temporal feature is hence to decide whether a node is influential based on the timings of the interactions and not only on the fixed count of interactions. More specifically, the method proposed here gives less importance to old interactions while giving more attention to the newer ones.

Each interaction is multiplied by a weight based on when it was created. Old interactions are given a lower score than the recent ones. All these values are then summed up to obtain the overall score for a particular node. This score acts as a weight signifying the importance of the node. In this case, this weight is stored in the respective node in the graph database and is then queried to be used as a feature in the classification.

In addition to this feature, this study uses four other features based on baseline metrics to represent the topology of the graph. The Katz metric is used since it was advocated by various studies as being one of the best indicators when it comes to predicting new links. Other simple metrics are used such as the common neighbours, shortest distance, and Jaccard's coefficient.

The sixth feature to describe is collated from the *friends* relationship in the graph database. This feature hence depicts whether two users are possibly online friends.

The experiment is run using R[1] programming language. Various packages are used to connect with the graph database and to provide various graph mining algorithms.

Data is first loaded from the graph database and stored as a graph in a main memory structure in R. Two subsets of this graph are taken to represent interactions during a specific period, and interactions created during a subsequent time period. As illustrated in Fig. 1, the first subgraph is used to return interactions in a specific time frame (i.e. t_1 to t_2), whereas the second subgraph consists of interactions which happened at a future time period (i.e. t_3 to t_4 where $t_3 > = (t_2 + 1)$).

[1] https://www.r-project.org/.

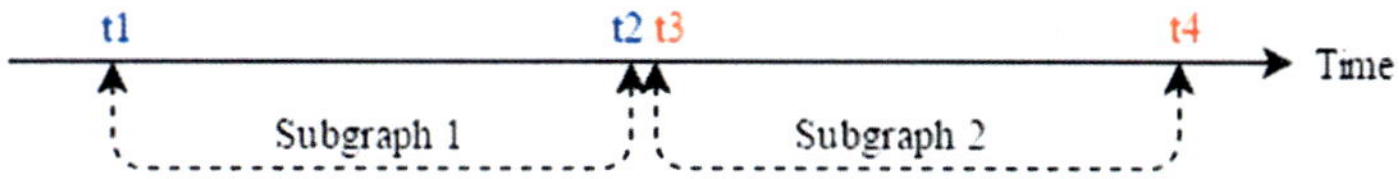

Fig. 1. Subgraphs created based on time intervals.

The baseline methods are run over subgraph 1 since this is used to predict whether a link exists in subgraph 2. Each method returns a matrix with scores obtained for all pairwise combinations. These scores are created based on the equations outlined in Sect. 2.1. Four matrices are available, each representing a metric; Katz, common neighbours, shortest distance, and Jaccard's coefficient. These scores together with all the other features are stored in a data structure where each row represents a pair of nodes and all the features corresponding to them.

The weights calculated by the temporal method are used as another feature which feeds into the classifier. Since this data structure represent the interaction between a pair of nodes, each record has two temporal weights; one for each node. The weight for the two nodes could be used separately; meaning that an additional feature is introduced. This is suitable if the direction of the link is considered and hence the weight representing the source node is separate from the weight of the target node. This, however, is not required in this research as the direction is not considered in its predictions. The experiment here is only interested in predicting whether a link formulates and not whom initiates a connection. Therefore, the separation of the two weights is not required. Consequently, the weights of the two nodes could be aggregated by means of summation or otherwise, as long as it signifies the importance of the two nodes.

The feature describing whether the pair of nodes are friends, is simply a Boolean value which is set to 1 if the two persons are online friends.

Table 2. Data structure used to build the classification model.

ID1	ID2	CN	KA	SD	JC	TB	FR	Class
9	11	8.1	7.2	9.1	13.6	7.5	1	1
1	2	7.5	0.3	10.6	6.4	15.2	0	1
15	1	0.2	11.0	4.2	8.2	8.9	1	0
20	3	9.6	3.9	13.1	0.9	9.7	1	1
7	17	2.5	6.1	8.6	11.3	0.9	0	0
19	4	3.4	15.2	3.9	9.1	13.4	0	1
11	9	7.6	5.9	11.2	0.6	1.6	1	0

In addition to these features, the data structure also includes the class of the classification. The class indicates whether a link actually formulated in the future. This value is a Boolean variable signifying if a link was actually created and it is captured from the interactions found in subgraph 2. An example of this data structure is shown in Table 2 where it displays its construction. The ID1 and ID2 columns correspond to the IDs of the persons. The next four columns indicate scores obtained from the four baseline metrics where CN is the common neighbours, KA is the Katz metric, SD is the shortest distance, and JC is Jaccard's Coefficient. The column TB (time-based) describe the temporal feature and the FR column shows whether the two persons are online friends. Finally, the last column is the label class which indicates whether a link is formulated in a future time interval.

The data structure is divided into two; one used to train the classification model, and the other for testing purposes. The partitioning for the data volumes listed in Table 1 is set to 60:40 for the training set and the test set respectively.

The classifier used in this study is the SVM with a radial basis function (RBF) kernel. The testing set is fed to the SVM so that a model could be built based on the features and the resulting class. The model is built by finding a hyperplane which best segregates the two classes based on the scores of the features. Additionally, the SVM could be tweaked by setting different parameters such as cost and gamma which in turn affect how the model is built. A method is used to tune the SVM so that it returns the cost and gamma values which obtains the best results. In this case the values for the cost and gamma were 100 and 2 respectively, and the model is built based on these values.

The testing set is then used on the model to predict link creation. These predictions are then compared to the actual values found in the label class in the data structure, and the intersection between the two constitutes the accuracy of the predictions.

3.3 Performance Measures

The results obtained are evaluated using various performance measures. These measures are based on the number of true positive predictions (TP), false positive predictions (FP), true negative predictions (TN), and false negative predictions (FN). These measures could be interpreted in many ways such as ratings, and percentages. The true positives and the false positives are also plotted against each other sometimes to produce an ROC (receiver operating characteristic) graph. The area under the curve in this graph gives the AUC score which is used in this study.

Other measures used include accuracy, precision, and recall. The accuracy is calculated by adding the true positives with the true negatives and dividing the answer by the total number of predictions.

$$Accuracy = \frac{TP + TN}{TP + TN + FP + FN} \tag{5}$$

The precision (also referred to as the positive prediction value) and recall measures are normally presented together as they are both based on the relevance of the results. The precision looks at the true positive predictions over the total positive predictions, whereas the recall looks at the true positive predictions over the actual positives.

$$Precision = \frac{TP}{TP + FP} \tag{6}$$

$$Recall = \frac{TP}{TP + FN} \tag{7}$$

The negative prediction value is the opposite of the precision measure. This is calculated by the equation below.

$$Negative\ prediction\ value = \frac{TN}{TN + FN} \quad (8)$$

These measures are used on the results obtained in this experiment. These are outlined in the next section.

4 Results and Evaluation

The experiment was run on the social network created from our dataset. The results correspond to the likeliness of two users checking-in at the same location and day.

The performance measures outlined earlier were adopted to analyse how the classification performed when compared to the actual data. These performance measures were used on both the training set and the test set. The former is used to outline how the classifier performed in terms of how well the classes were segmented. High performance in such results indicate that the model built is representable of how data items should be classified and therefore correct predictions are more likely to be achieved. The test set is then used on the same model to depict how accurate the predictions are.

The results obtained are reproduced in Table 3 and evidently, the majority of results in for training are better than for testing. This was expected since the model was built on the data found in the training set and hence the predictions are more precise.

Table 3. Results obtained from the training and test set.

Set	Accuracy %	AUC	Recall %	Positive prediction value (precision)	Negative prediction value
Train	93.4	0.646	38.4	0.970	0.932
Test	94.0	0.505	38.0	0.044	0.967

However, both results yielded accuracies higher than 90% which means that the overall predictions were correct (both negative and positive). The AUC score for the train set is higher than the one obtained in the test set. Having said that, both scores are higher than 0.5 which means that they both performed well in general.

The recall values show that 38.4% and 38% of the actual links formulated were correctly predicted. A significant discrepancy is seen in the precision value (positive prediction value). This shows that the results in the test set predicted a higher number of positive links which were not actually formulated (false positives). This is evident since the accuracy and recall for both data sets yielded similar values and the main difference is only seen in the precision value. Obtaining a low precision value was expected due to the challenges posed by the topology of social networks, however, one can also argue that this value can also indicate a number of missing links in the graph. These missing links could have been omitted from the graph, or formed at a later timeframe than the one chosen in this experiment.

The negative prediction value shows that a good number of the negative predictions were also correctly predicted. Once again predicting less positive links than negative ones was also expected due to the challenge posed by the sparsity of social networks.

5 Conclusions

The study outlined here shows a new perspective of how spatial data on social interactions could be used to build social networks. The experiment made use of traditional graph features together with a time based feature to predict the likelihood of two person checking-in at the same place over on the same day.

Knowing such information beforehand can be useful in various applications. The results obtained in this study showed that even though the eventuality of persons being at the same place could be hard to infer, past information is beneficial to depict such predictions.

The graph model is also useful when analysing such data and patterns since the topology itself offers indicative measures that are useful when building these predictions. Additionally, considering temporal behaviour helped in the classification process.

This experiment showed interesting results which could be used in link predictions. This motivates the authors to explore how other features which consider the geographical distance between localities might help in achieving improved results. Having the data at hand and using the model introduced here allows for further research.

Furthermore, this research could be used to build models that rather than just predicting that two persons will be at the same place, the model will also try to infer the location where such interaction will take place.

References

1. Al Hasan, M., Chaoji, V., Salem, S., Zaki, M.: Link prediction using supervised learning. In: Proceedings of SDM 06 workshop on Link Analysis, Counterterrorism and Security (2006)
2. Liben-Nowell, D., Kleinberg, J.: The link prediction problem for social networks. J. Am. Soc. Inf. Sci. Technol. **58**(7), 1019–1031 (2007)
3. Dashun, W., Pedreschi, D., Song, C., Giannotti, F., Barabasi, A.L.: Human mobility, social ties, and link prediction. In: Proceedings of the 17th ACM SIGKDD International Conference on Knowledge Discovery and Data Mining (2011)
4. Chen, H., Li, X., Huang, Z.: Link prediction approach to collaborative filtering. In: Proceedings of the 5th ACM/IEEE-CS Joint Conference on Digital Libraries (2005)
5. Mallek, S.: Evidential link prediction in social networks based on structural. J. Comput. Sci. **30**, 98–107 (2019)
6. Getoor, L., Diehl, C.P.: Link mining: a survey. ACM SIGKDD Explor. Newsl. **7**(2), 3–12 (2005)
7. Lichtenwalter, R.N., Lussier, J.T., Chawla, N.V.: New perspectives and methods in link prediction. In: Proceedings of the 16th ACM SIGKDD International Conference on Knowledge (2010)

8. Leskovec, J., Backstrom, L.: Supervised random walks: predicting and recommending links in social networks. In: Proceedings of the fourth ACM International Conference on Web Search and Data Mining (2011)
9. Potgieter, A., April, K.A., Cooke, R.J.E.: Temporality in link prediction: understanding social complexity. Emergence: Complexity Organ. **11**(1), 69–83 (2009)
10. Munasinghe, L.: Time-aware methods for Link Prediction in Social Networks, Department of Informatics School of Multidisciplinary Sciences the Graduate University for Advanced Studies (2013)
11. Dong, L., Li, Y., Yin, H., Le, H., Rui, M.: The algorithm of link prediction on social network. Math. Prob. Eng. **4** (2013)
12. Gupta, N., Singh, A.: A novel strategy for link prediction in social networks. In: Proceedings of the 2014 CoNEXT on Student Workshop ACM, pp.10–14 (2014)
13. Scellato, S., Noulas, A., Cecilia, M.: Exploiting place features in link prediction on location-based social networks. In: Proceedings of the 17th ACM SIGKDD International Conference on Knowledge Discovery and Data Mining, pp. 1046–1054. San Diego, CA (2011)
14. Ayoub, J., Lotfi, D., El Marraki, M.: Accurate link prediction method based on path length between a pair of unlinked nodes and their degree. Soc. Netw. Anal. Min. **10** (2020)
15. Ichise, R., Rümmele, N., Werthner, H.: Exploring supervised methods for temporal link prediction in heterogeneous social networks. In: WWW 2015 Companion Proceedings of the 24th International Conference on World Wide Web Pages, pp. 1363–1368 (2015)
16. Ricardo, P., Soares, S., Prudencio, R.B.C.: Time series based link prediction. In: The 2012 International Joint Conference on Neural Networks (IJCNN), Brisbane, QLD, Australia (2012)
17. Hajibagheri, A., Sukthankar, G., La, K.: Leveraging Network Dynamics for Improved Link Prediction, Social, Cultural, and Behavioral Modeling. SBP-BRiMS 2016. Lecture Notes in Computer Science. Springer (2016). https://doi.org/10.1007/978-3-319-39931-7_14
18. Leskovec, J., Krevl, A.: SNAP Datasets: Stanford Large Network Dataset Collection, June 2014. http://snap.stanford.edu

Comparing Classification Algorithms on Predicting Loans

Krishanu Agarwal(✉), Mohit Jain, and Ashok Kumawat

Department of Mechatronics Engineering, Manipal University Jaipur, Jaipur, Rajasthan, India
krishanu.179403029@muj.manipal.edu, {mohit.jain, ashokkumar.kumawat}@jaipur.manipal.edu

Abstract. Bank loans are becoming difficult to manage with increase in purchasers in the 21st century. This rate of increase in the loans, including student loans and corporate loans is a demand for more manpower. In addition, there is huge requirement for spotting persons who are capable of being potential buyers of personal loans. The information gained by analysing the data collected from banks which is blended through different algorithms while applying various statistical calculations proves liable in the prediction of probable customers. Machine learning is now being used to apply such tedious calculation and to provide insights for further predictive study for loan clearance and risks involved with it. The following attempt is on comparing the behaviour of machine learning algorithms: logistic regression, random forest, decision tree, support vector classifier, k-nearest neighbours, in terms of accuracy, precision and recall, true positive and false positive rate as the model predicts the person's capability of purchasing personal loan. It is seen that random forest algorithm performed the best with 98% accuracy and k-nearest neighbours performed the poorest with 90% accuracy. To validate these terms the task is carried out on Google Colab using Python.

Keywords: Machine learning · Logistic regression · Random forest · Decision tree · Support vector classifier · k-nearest neighbours

1 Introduction

Expecting over 69% adults – 3.8 billion people worldwide with bank accounts [1] and with the ever-increasing loans given like student loans, property loans, business loans, the management of these loans has become a burden. Therefore, feasible method for accurate prediction of the various bank related issues is profoundly critical. The pertaining data, collected by national banks, is humongous and is subjected to noise and outliers. The visualizing of these datasets proves to be very difficult for humans to visualize in view of the size. This task becomes easier and faster with the help of machine learning tools.

There are two types of machine learning: supervised learning (data is labelled) and unsupervised learning (data is not labelled). Classification comes under supervised learning along with regression. Regression is done when the target value is continuous (like house prices), whereas classification is done when the data is categorical (like gender-male or female). Standard techniques like logistic regression, widely used for binary

L. Garg et al. (Eds.): ISMS 2020, LNNS 303, pp. 240–249, 2022.
https://doi.org/10.1007/978-3-030-86223-7_21

Author Index

L. Garg et al. (Eds.): ISMS 2020, LNNS 303, pp. 251–252, 2022.
https://doi.org/10.1007/978-3-030-86223-7